HOW TO DRAW EVERYTHING FOR KIDS

101 Simple Drawings with Step-by-Step Instructions for Transportation, Sports, Superheroes, Animals, and Other Fun Items!

"Welcome to "How to Draw Everything for Kids"! In this book, you'll find easy step-by-step guides for drawing all sorts of cool things like vehicles, superheroes, animals, and more!

So grab your pencils and start your drawing adventure!"

THIS BOOK BELONGS TO:
Write your name here
and make this book your own!

TIPS AND TOOLS

To get the most out of this drawing book, here are some tips and tools you'll need:

Pencil:
Use a pencil to sketch your drawings. It's okay to make mistakes – you can always erase!

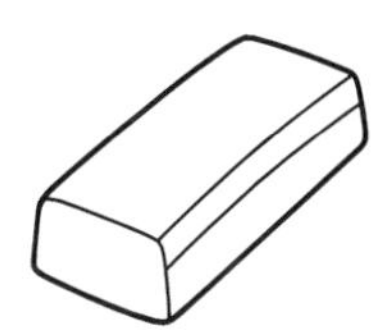

Eraser:
A good eraser helps you fix any mistakes and keep your drawings clean.

Colored Pencils or Crayons:
Add color to your drawings to make them come to life!

Sketchbook or Drawing Paper:
Practice on extra paper before drawing in the book.

"Remember, practice makes perfect. Keep trying and have fun!"

INSTRUCTIONS

This book will help you learn to draw all sorts of fun things, step by step.

Here's how it works:

1. **Follow the Steps**:

Each drawing is broken down into simple steps. Start with step 1 and follow the numbers until you complete the drawing. New lines for each step are shown in black, and lines from previous steps are shown in gray.

2. **Practice**:

Use the "Practice" area on each page to try drawing on your own. Don't worry if it's not perfect - practice makes you better!

3. **Have Fun**:

Drawing should be fun! Feel free to add your own touches and make each drawing your own.

- 08 • Racing car
- 09 • Motorcycle
- 10 • Airplane
- 11 • Helicopter
- 12 • Bicycle
- 13 • Truck
- 14 • Train
- 15 • Boat
- 16 • Ship
- 17 • Rocket
- 18 • Flying saucer
- 19 • Scooter
- 20 • Submarine
- 21 • Fire truck
- 22 • Police car
- 23 • Tractor
- 24 • Limousine
- 25 • Bus
- 26 • Bulldozer
- 27 • Electric scooter
- 28 • Paraglider
- 29 • Jet ski
- 30 • Jeep
- 31 • Tank
- 32 • Fighter jet
- 33 • Football
- 34 • Basketball
- 35 • Hockey puck
- 36 • Baseball bat
- 37 • Jersey
- 38 • Sports sneakers
- 39 • Tennis racket
- 40 • Boxing gloves
- 41 • Rugby ball
- 42 • Football goal
- 43 • Cap
- 44 • Foosball table
- 45 • Championship cup
- 46 • Sports medal
- 47 • Snowboard
- 48 • Rollerblades
- 49 • Frisbee
- 50 • Skateboard

Superheroes and Fantasy Characters & Animals

51 • Superhero
52 • Knight in armor
53 • Ninja
54 • Wizard
55 • Viking
56 • Alien
57 • Dragon
58 • Transformer
59 • Drone
60 • Robot dinosaur
61 • Griffin
62 • Pegasus
63 • Giant spider
64 • Centaur
65 • Pumpkin
66 • Ghost
67 • Pirate
68 • Tiger
69 • Wolf
70 • Elephant
71 • Shark
72 • Bear
73 • Crocodile
74 • Eagle
75 • Dog
76 • Cat
77 • Parrot
78 • Penguin
79 • Jellyfish
80 • Sparkle the Little Fox
81 • Lion

Technology, Gadgets and Items & Food

82 • Gaming console
83 • Lightsaber
84 • Smartphone
85 • Laptop
86 • Virtual reality glasses
87 • Robot vacuum cleaner
88 • Headphones
89 • Camera
90 • Electric toothbrush
91 • Marker
92 • Water gun
93 • Globe
94 • Backpack
95 • Rocking chair
96 • Smartwatch

97 • Slice of pizza
98 • Burger
99 • Sandwich
100 • Hot dog
101 • French fries
102 • Nachos
103 • Lemonade
104 • Apple
105 • Cake
106 • Ice cream
107 • Cupcake
108 • Pancakes

Racing cars can go as fast as 200 miles per hour!

2

3

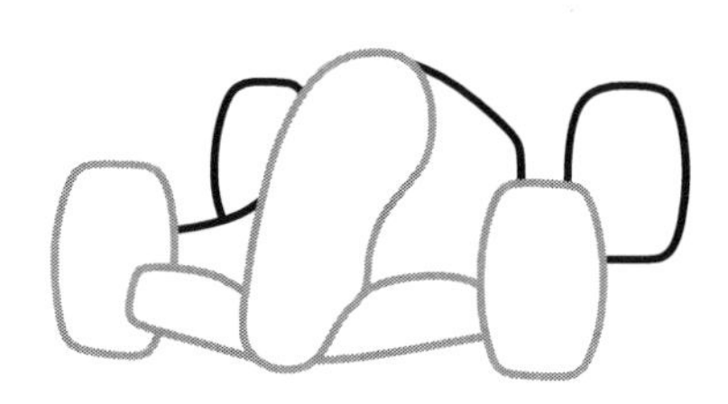

4

5

Now, try to draw

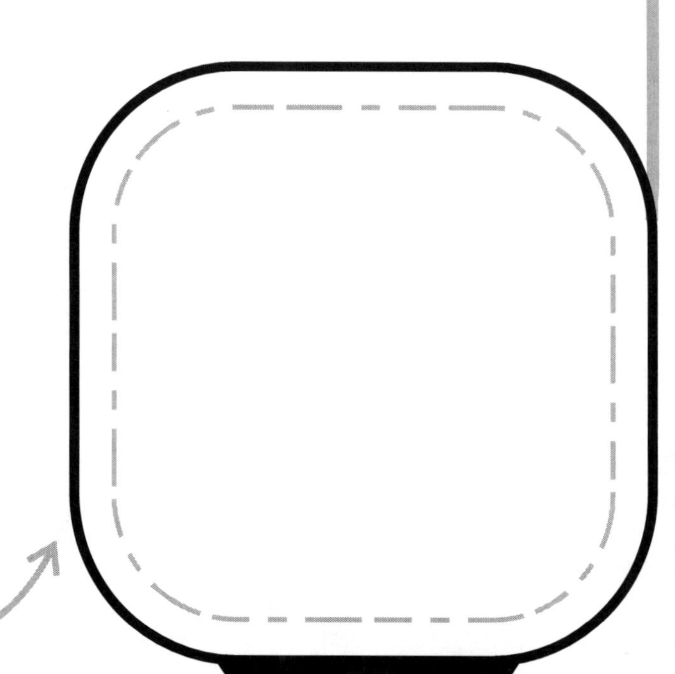

Motorcycles have two wheels and are very fast on the road

2

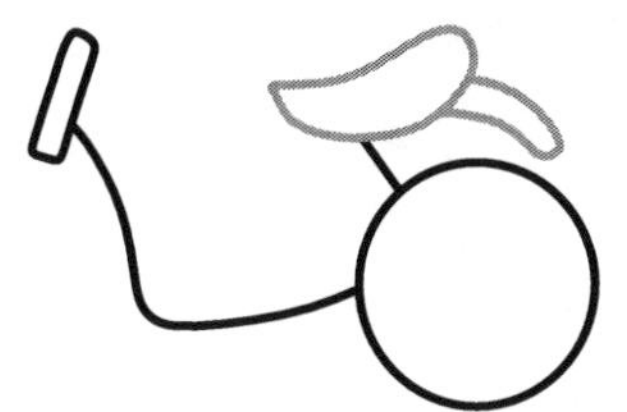

3

4

5

Now, try to draw

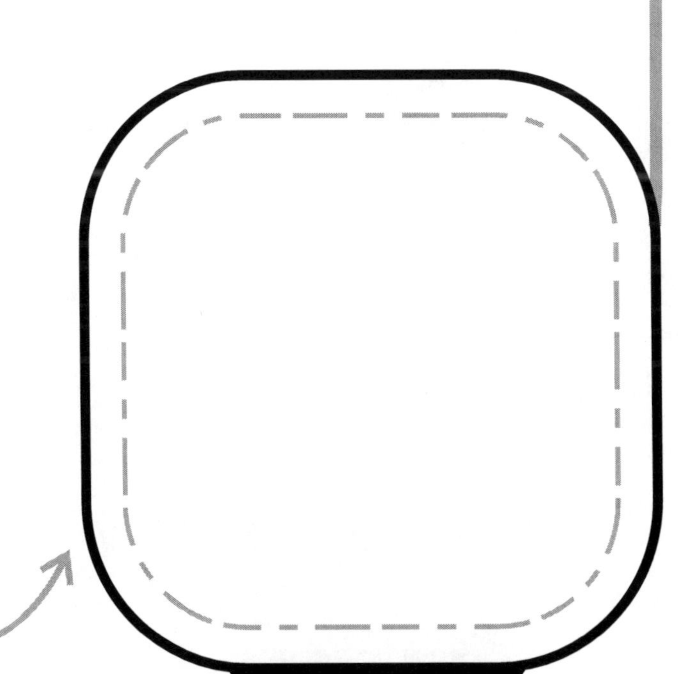

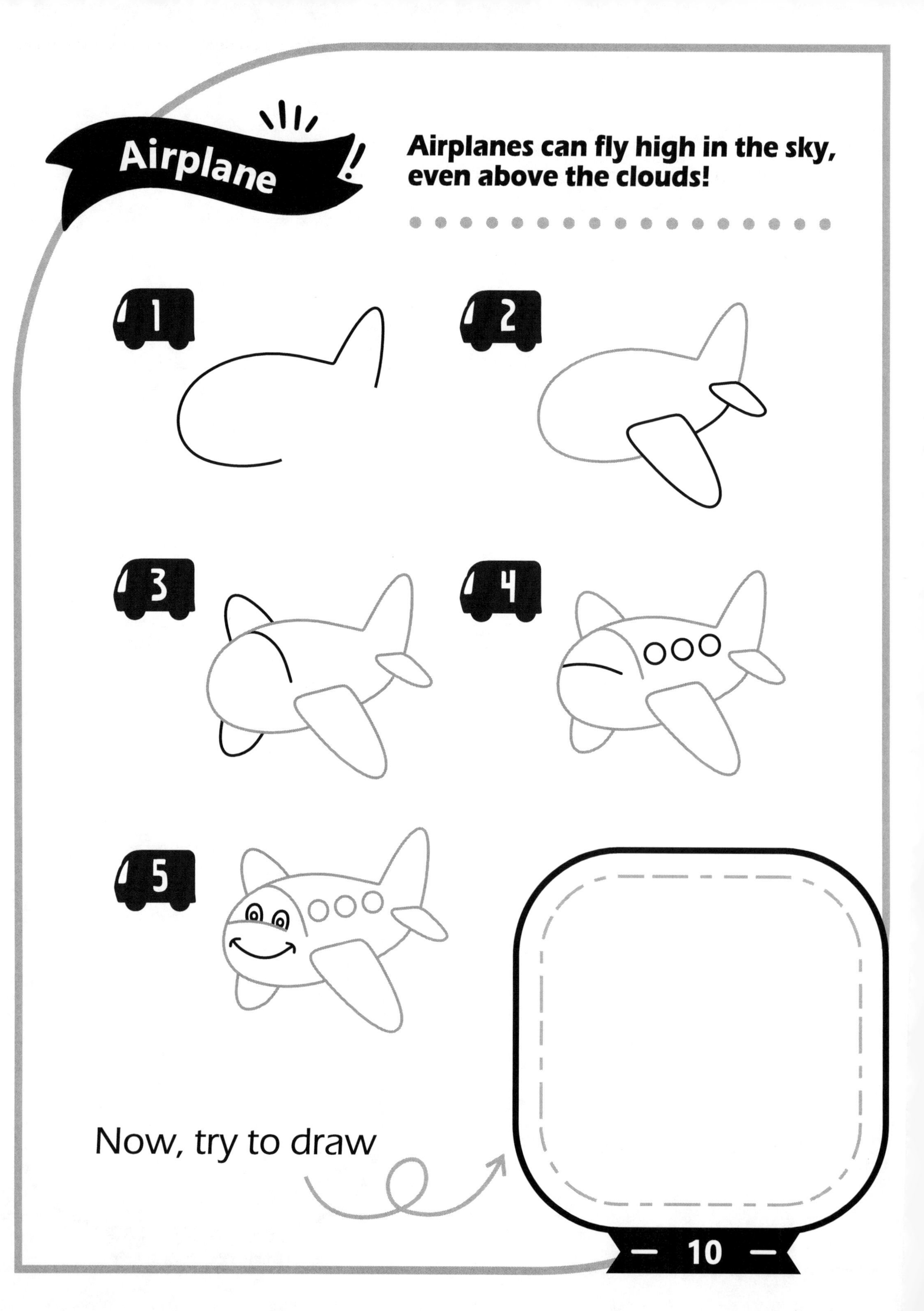
Airplane
Airplanes can fly high in the sky, even above the clouds!
1
2
3
4
5
Now, try to draw

Helicopter

Helicopters can fly straight up and down and even hover in place

1

2

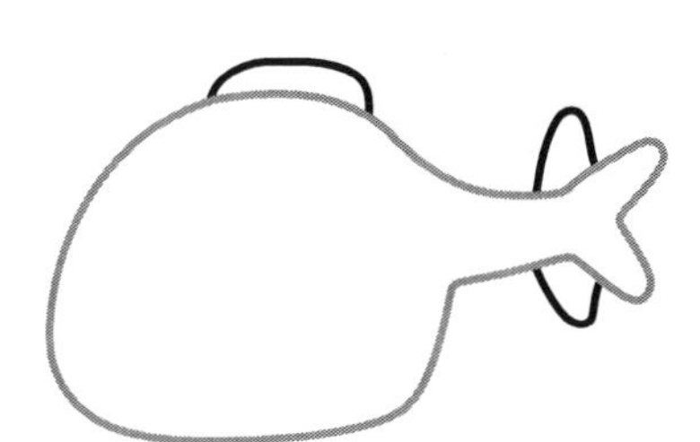

3

4

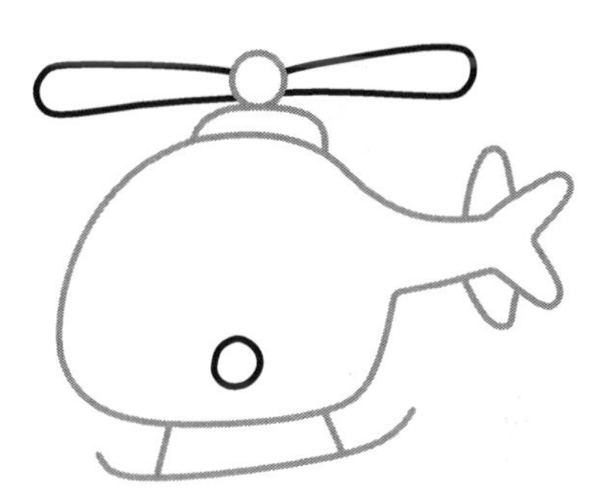

5

Now, try to draw

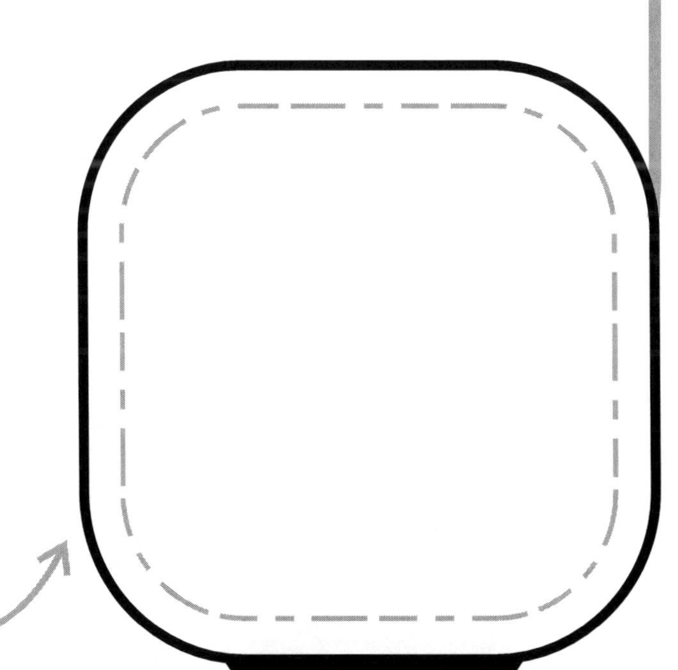

Riding a bicycle is great exercise and a fun way to explore!

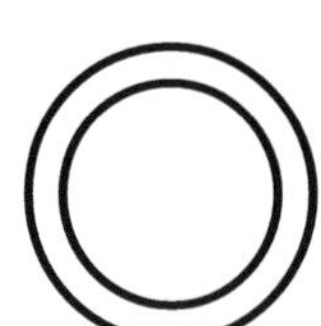

Now, try to draw

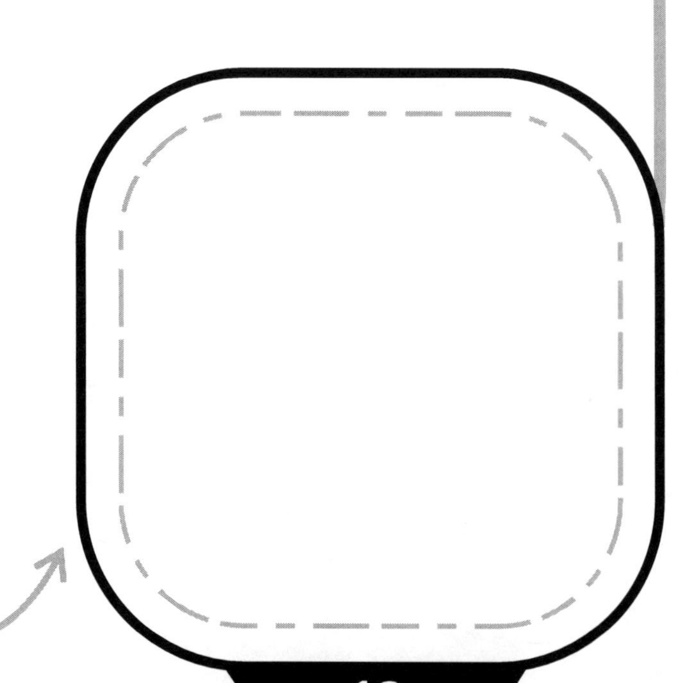

Truck
Trucks can carry really big loads, like furniture or food
1
2
3
4
5
Now, try to draw

Some trains can travel at speeds over 300 miles per hour!

2

5

Now, try to draw

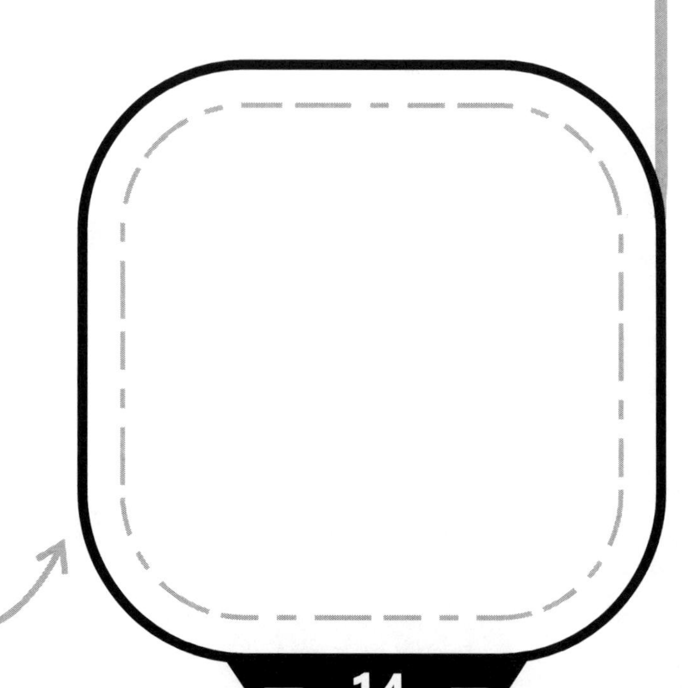

Boats float on water and can be used for fishing, travel, or fun!

Now, try to draw

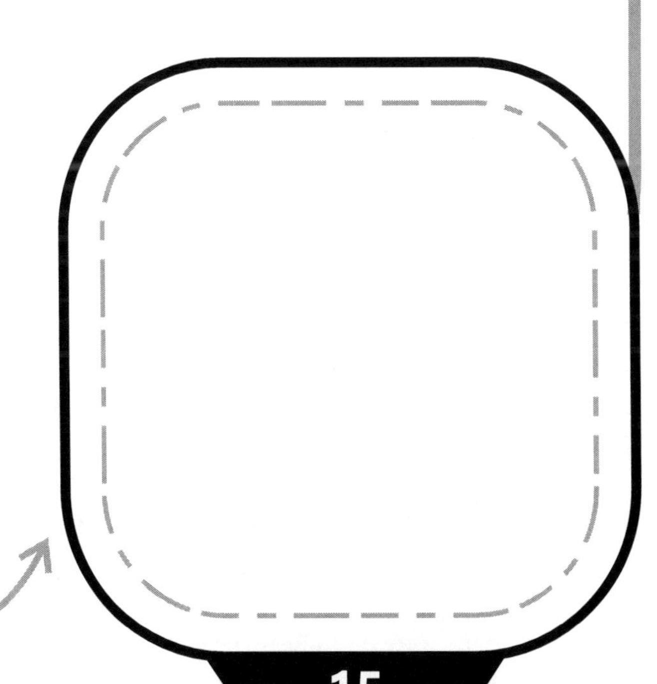

Ship
Ships are much bigger than boats and can carry hundreds of people!
1
2
3
4
5
Now, try to draw

Rocket

Rockets can travel into space, far beyond our planet!

1

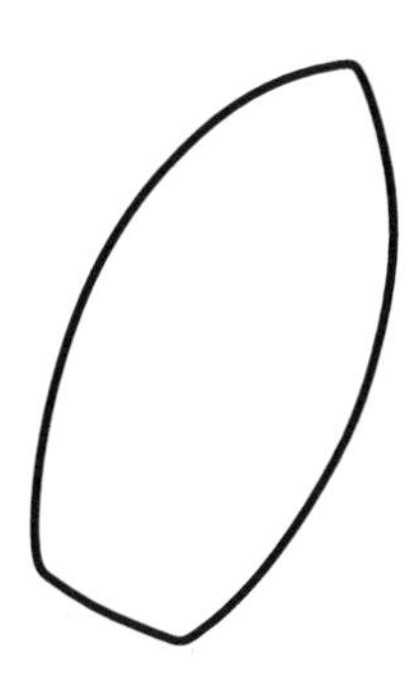

2

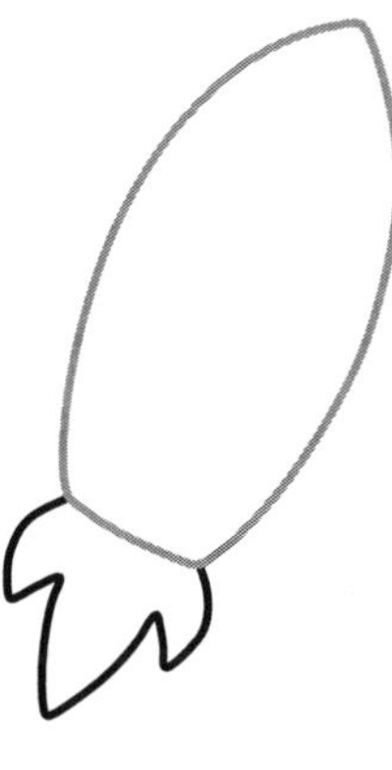

3

4

5

Now, try to draw

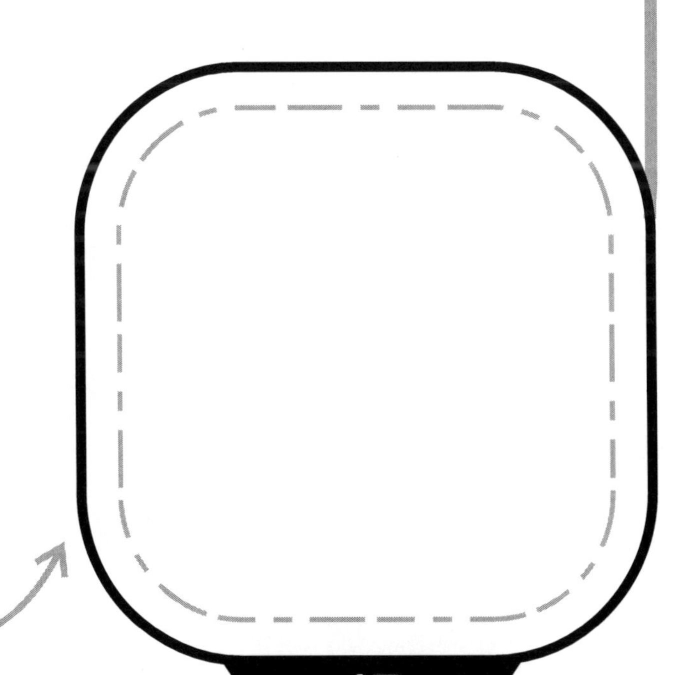

Flying saucers are fun to imagine in science fiction stories about aliens!

1

2

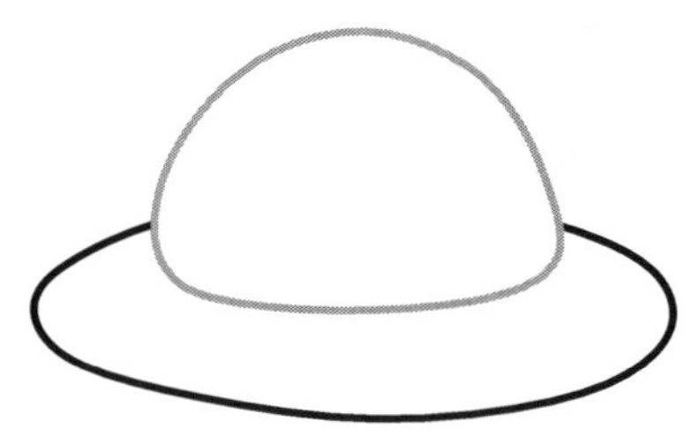

3

4

5

Now, try to draw

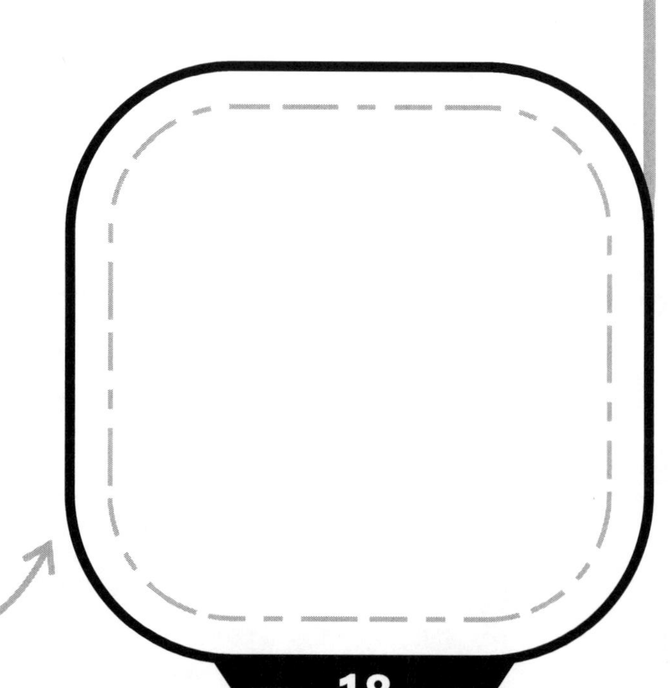

Scooter

Scooters are fun for zipping around the neighborhood

1

2

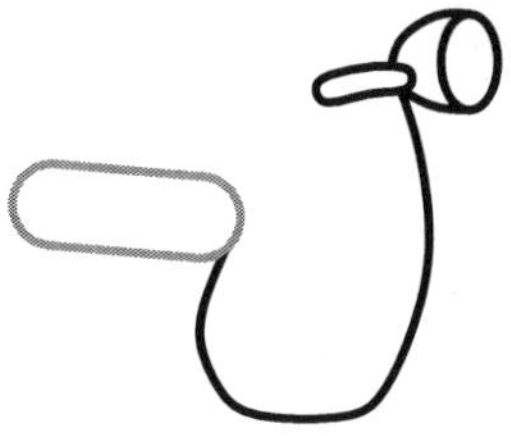

3

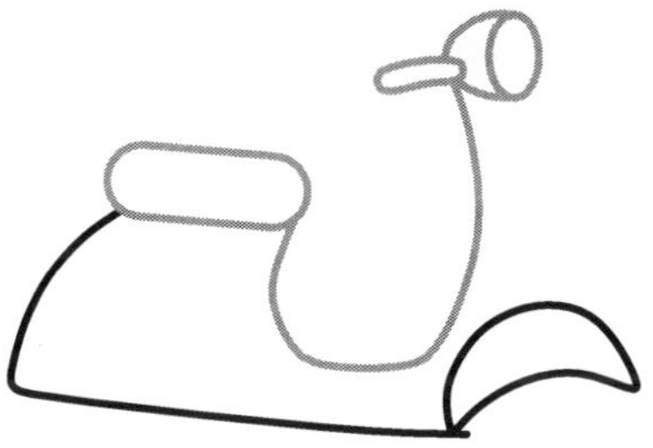

4

5

Now, try to draw

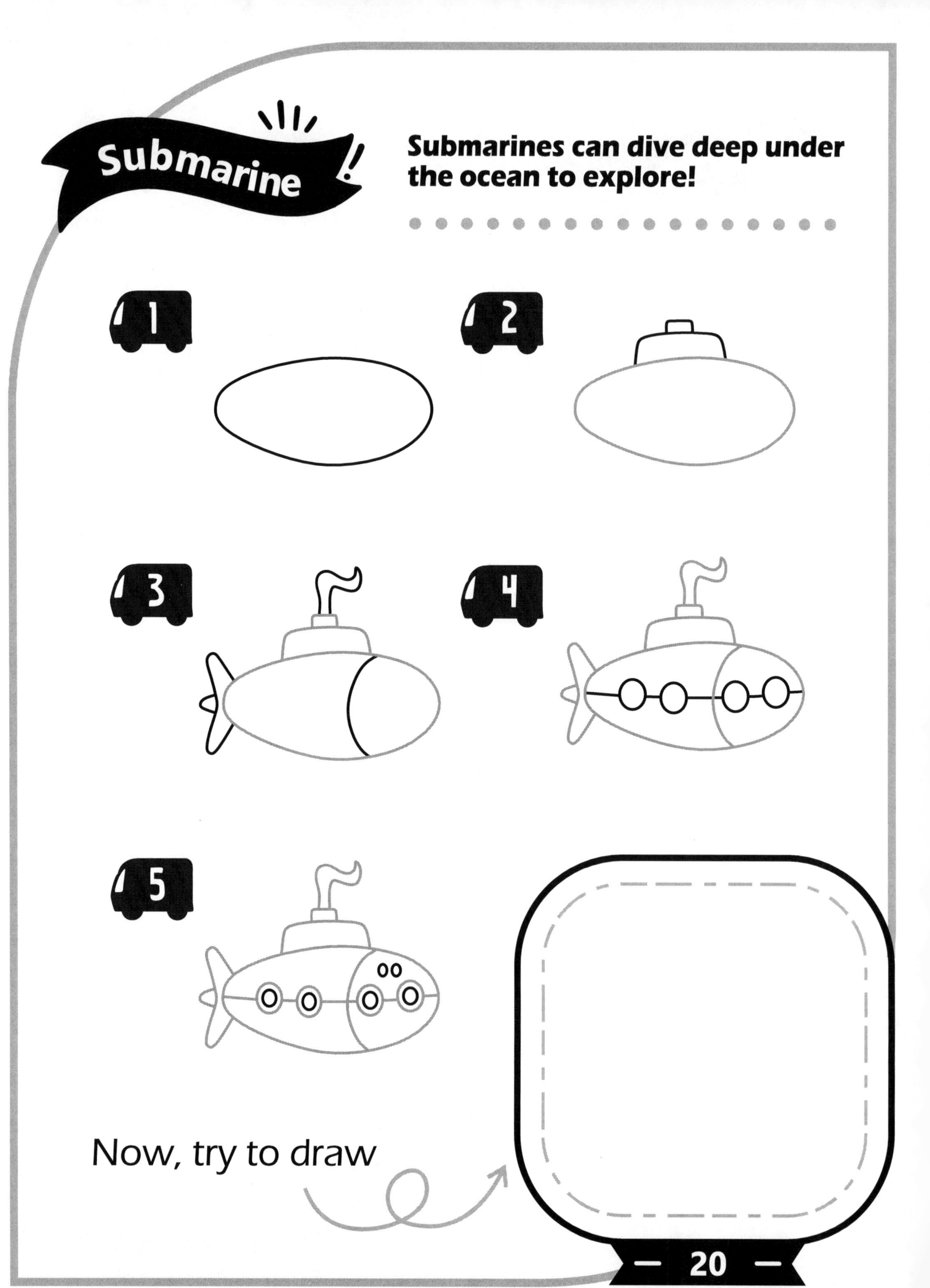
Submarine
Submarines can dive deep under the ocean to explore!
1
2
3
4
5
Now, try to draw

Fire truck

Fire trucks have ladders and hoses to help put out fires

1

2

3

4

5

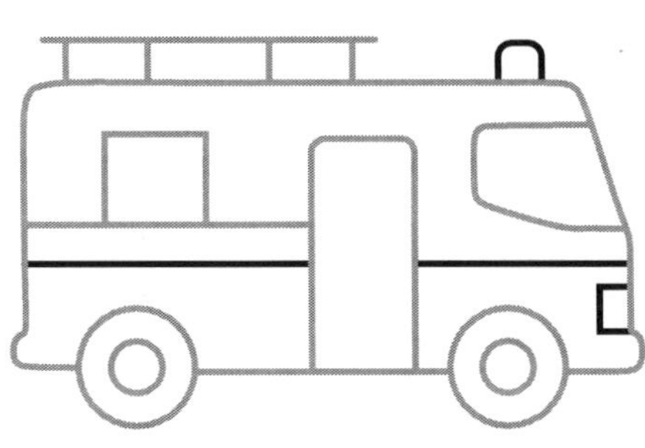

Now, try to draw

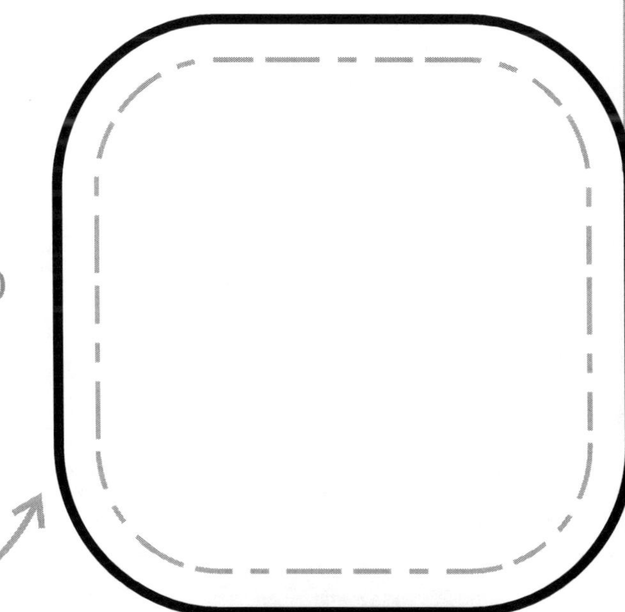

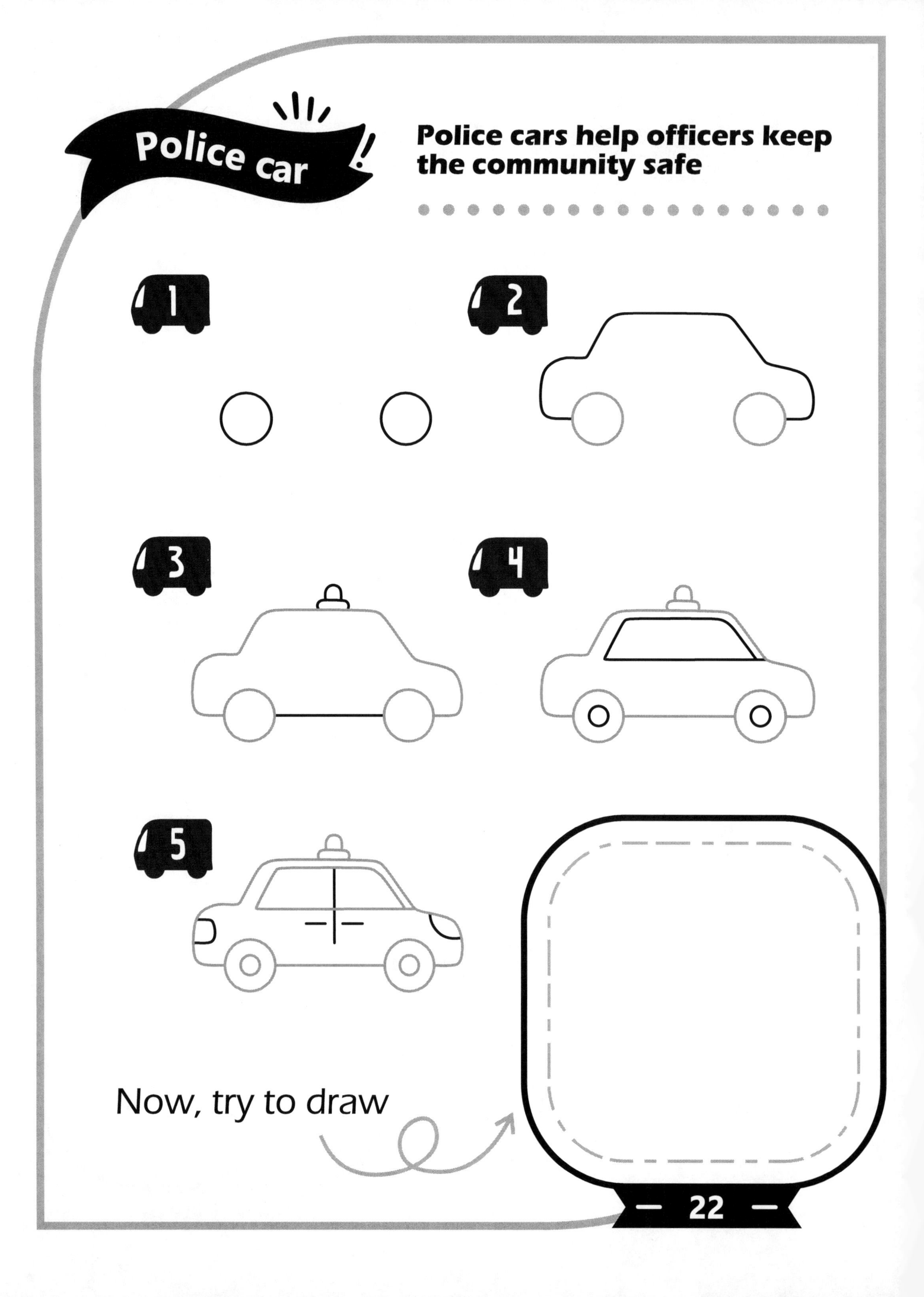
Police car
Police cars help officers keep the community safe
1
2
3
4
5
Now, try to draw

Tractor
Tractors are used on farms to plow fields and harvest crops
1
2
3
4
5
Now, try to draw

Limousine
Limousines are super long and often used for special events
1
2
3
4
5
Now, try to draw

Bus
Buses can carry lots of people around town or to school
1
2
3
4
5
Now, try to draw

Bulldozer!

Bulldozers have a big blade in front to move dirt and rocks

1

2

3

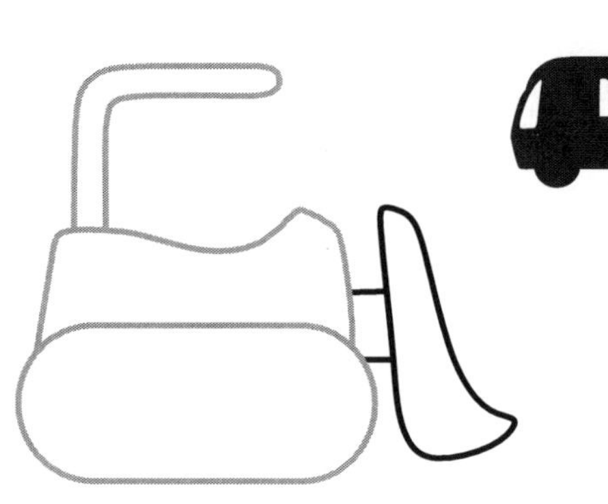

4

5

Now, try to draw

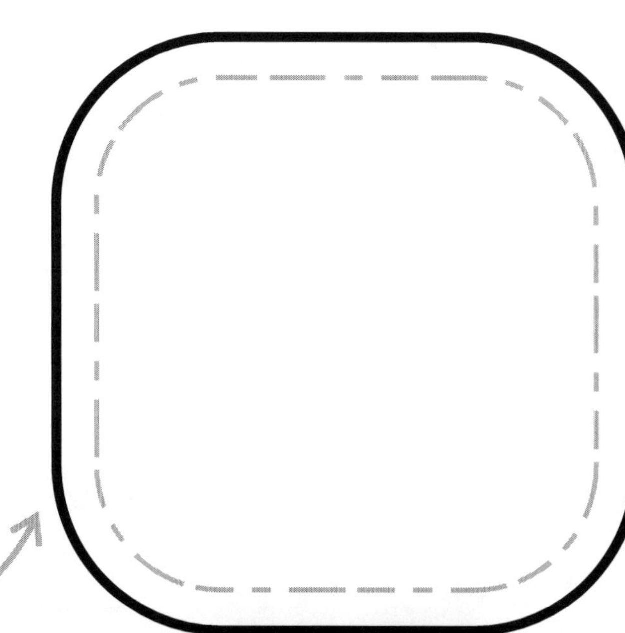

Electric scooters are powered by batteries and are fun and eco-friendly!

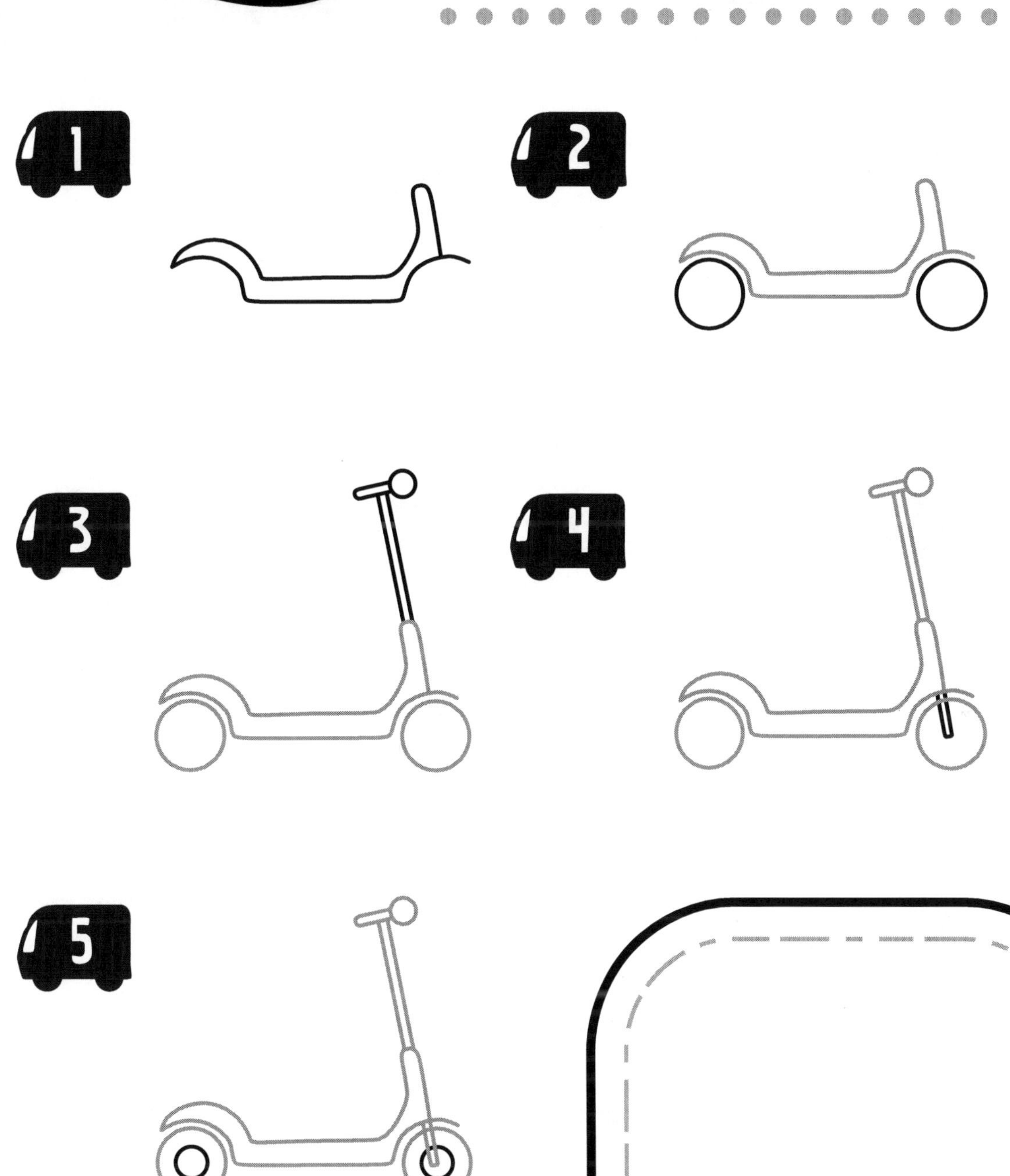

Now, try to draw

Paraglider!

Paragliders float through the air using a parachute-like wing

1

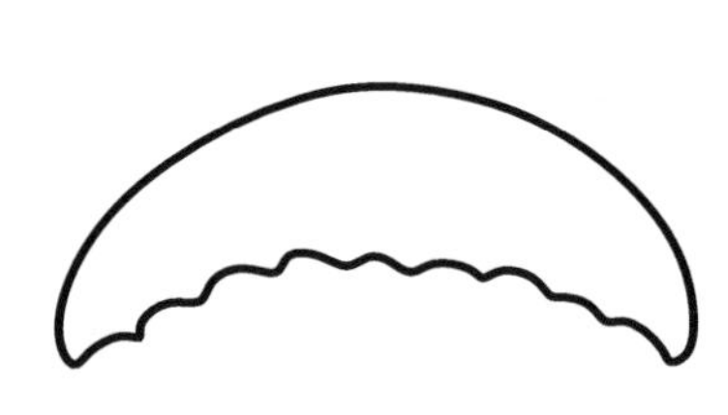

2

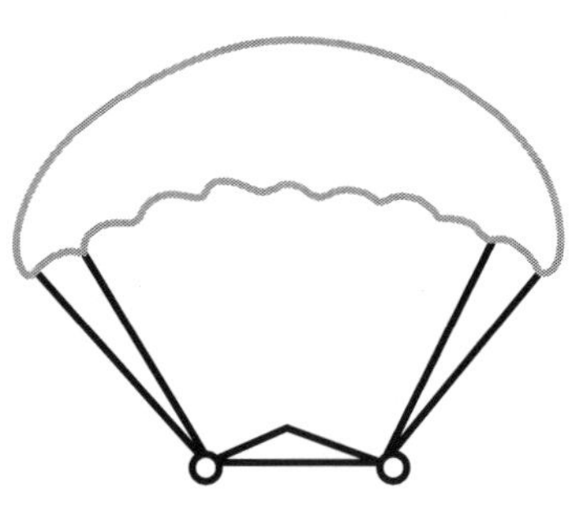

3

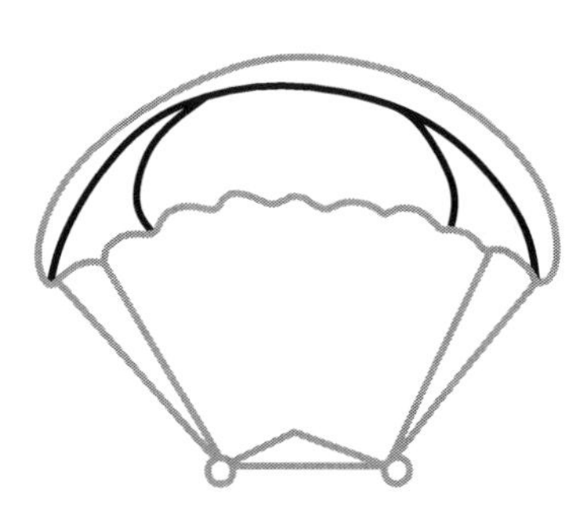

4

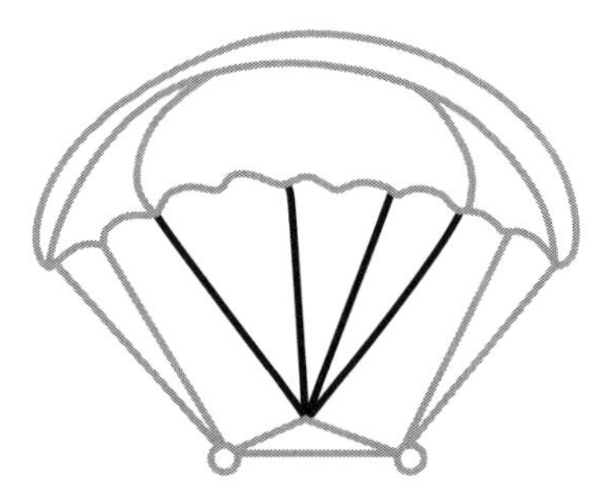

5

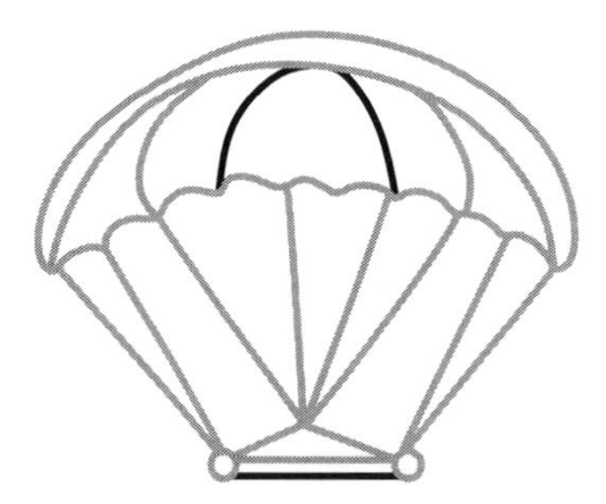

Now, try to draw

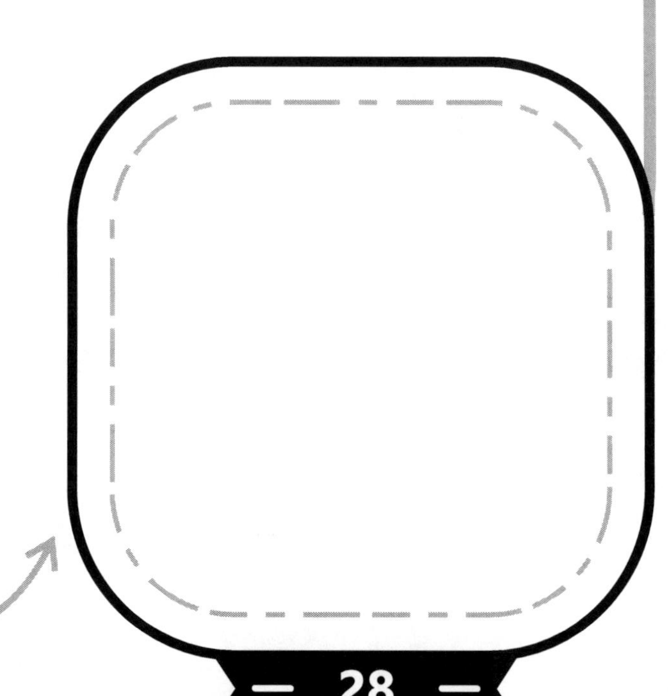

Jet ski
Jet skis zoom across the water, like motorcycles on the sea!
1
2
3
4
5
Now, try to draw

Jeep
Jeeps are great for off-road adventures in the mountains or desert
1
2
3
4
5
Now, try to draw

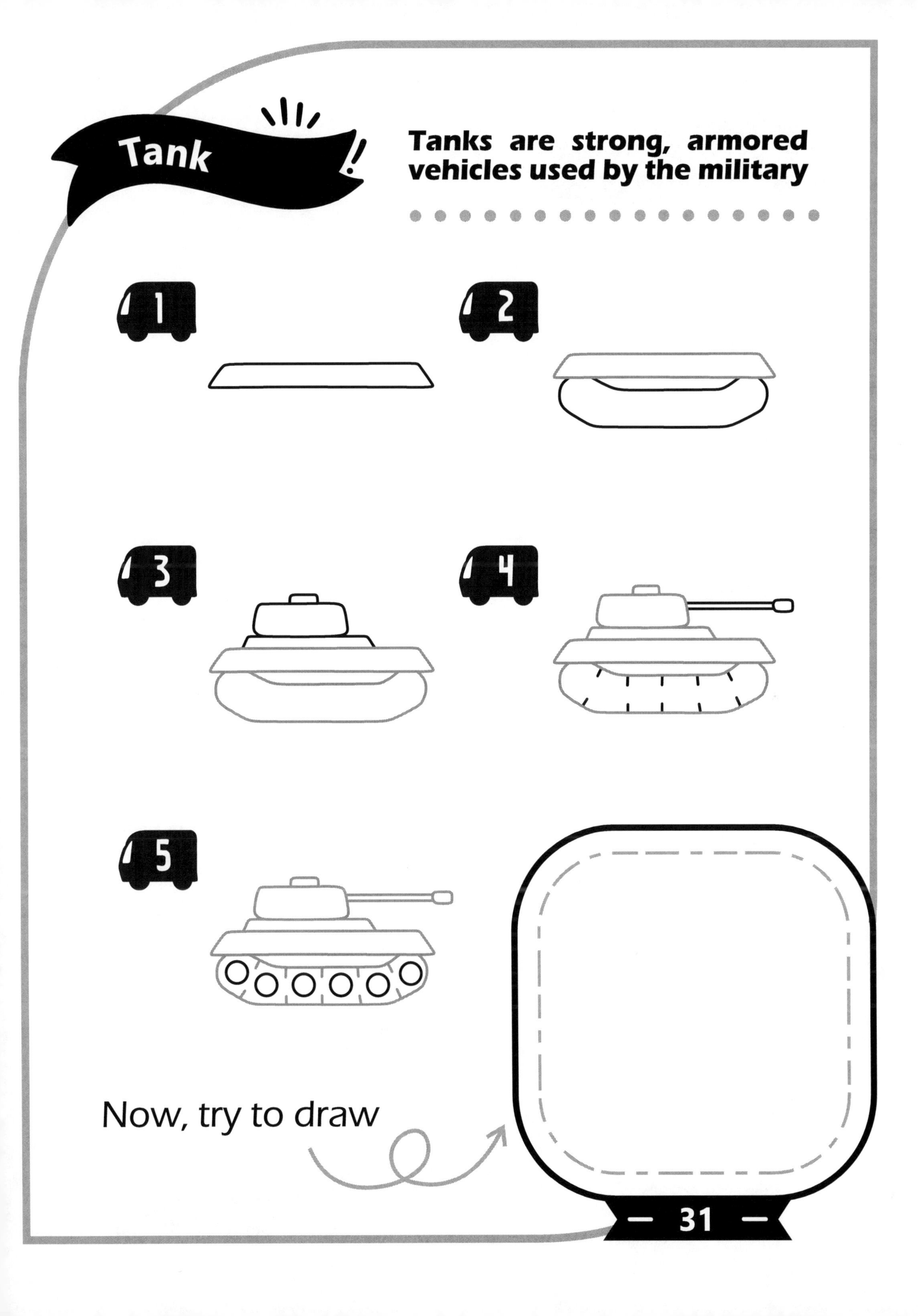
Tank
Tanks are strong, armored vehicles used by the military
1
2
3
4
5
Now, try to draw

Fighter jet

Fighter jets can fly faster than the speed of sound!

1

2

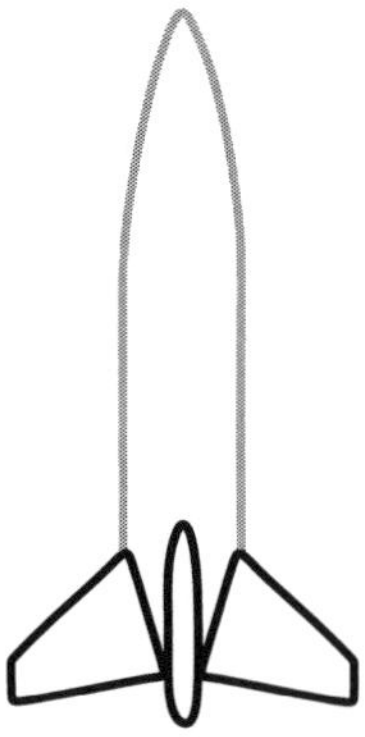

3

4

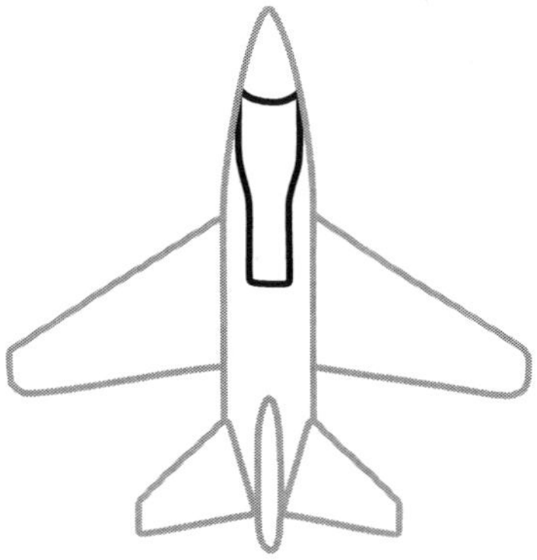

5

Now, try to draw

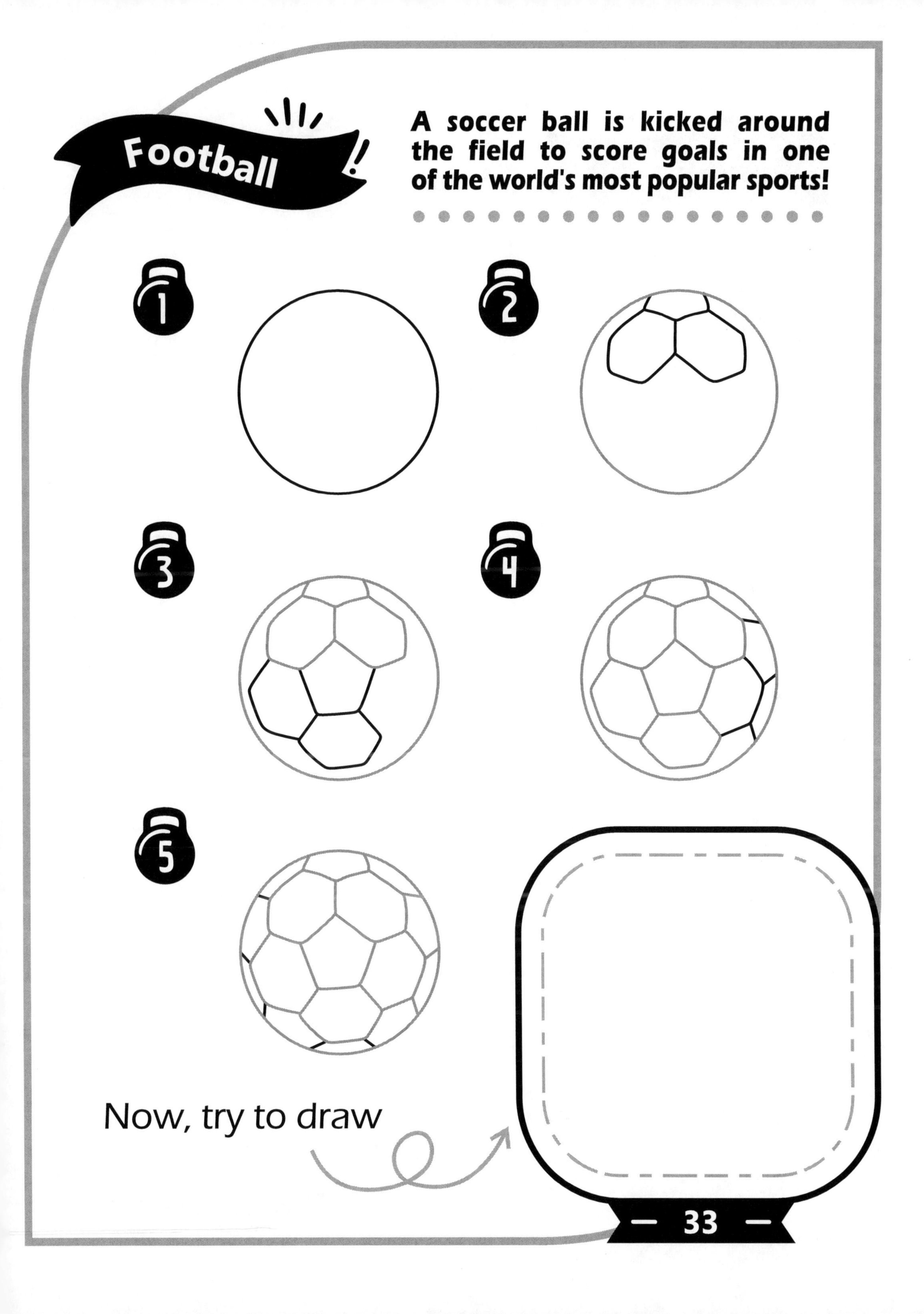
Football
A soccer ball is kicked around the field to score goals in one of the world's most popular sports!
1
2
3
4
5
Now, try to draw

Basketball

In basketball, players aim to shoot the ball into the hoop for points

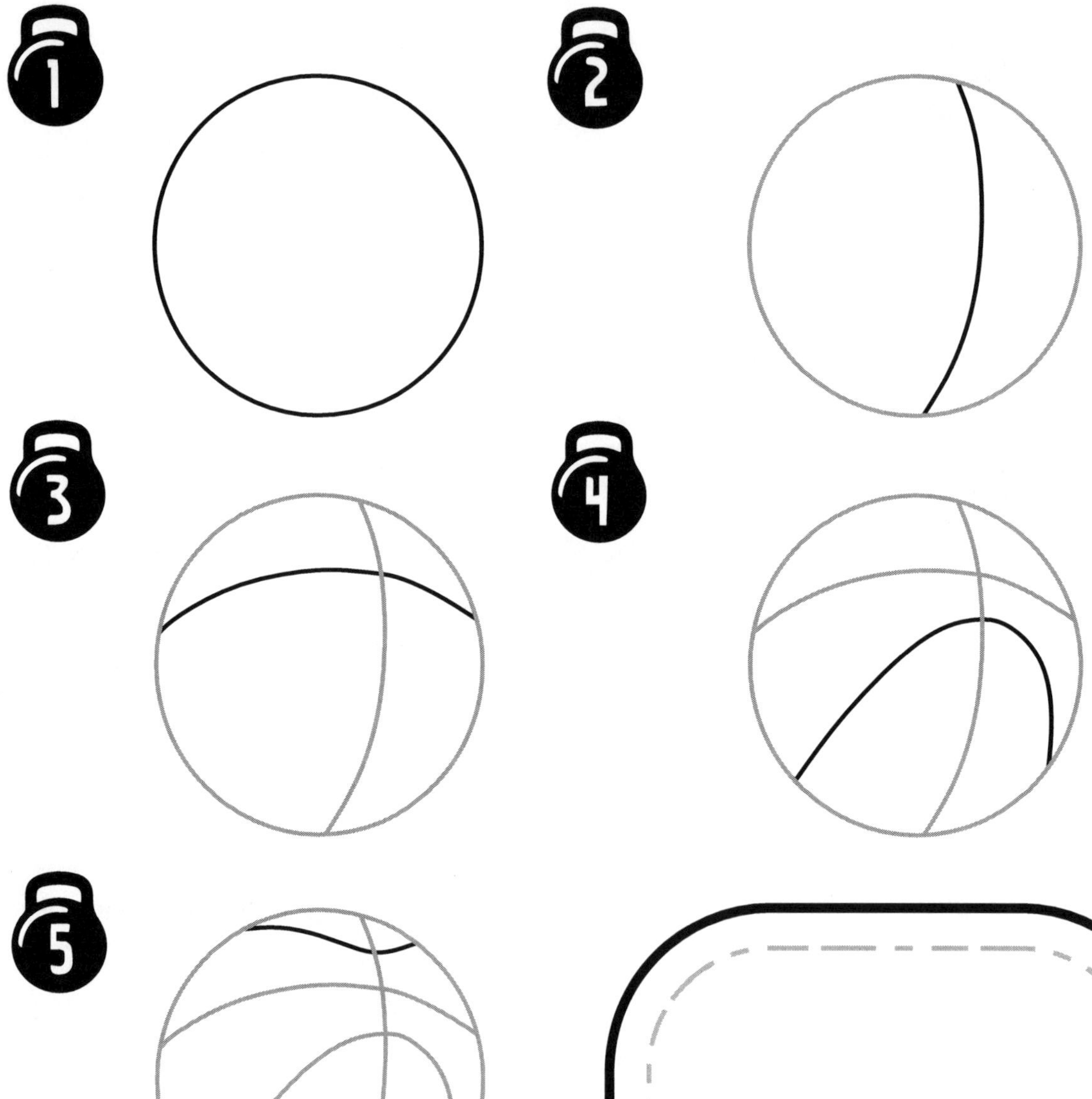

1

2

3

4

5

Now, try to draw

Hockey puck
A hockey puck is used in ice hockey, sliding across the ice at incredible speeds!
1
2
3
4
5
Now, try to draw

Baseball bat

A baseball bat is used to hit the ball and score home runs in baseball

1

2

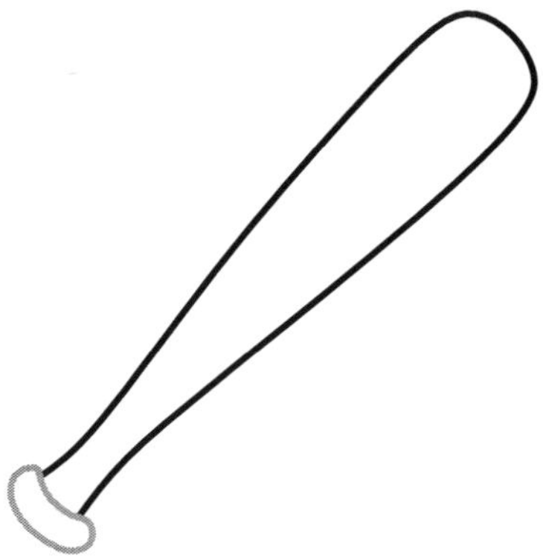

3

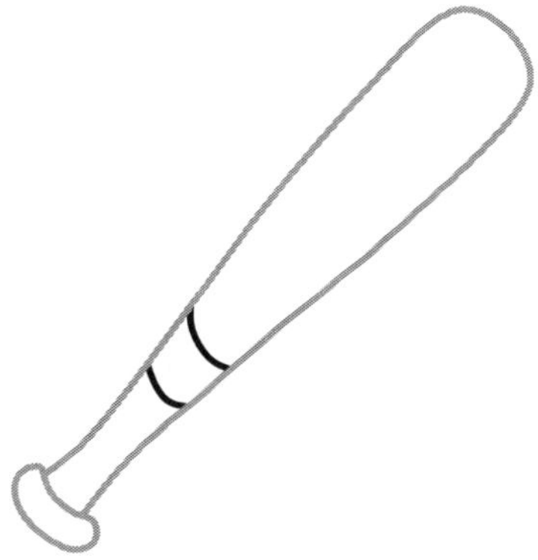

4

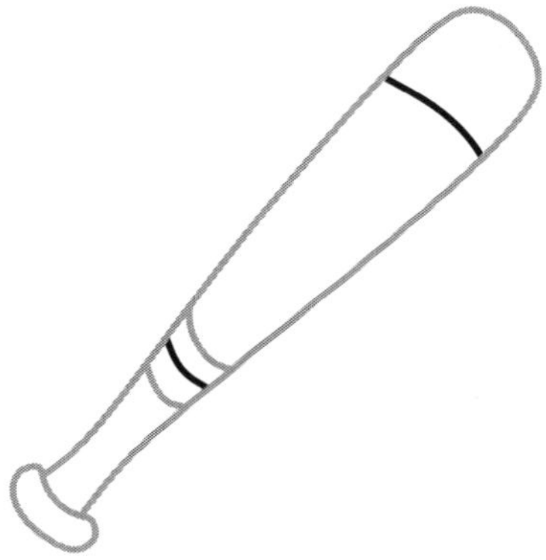

5

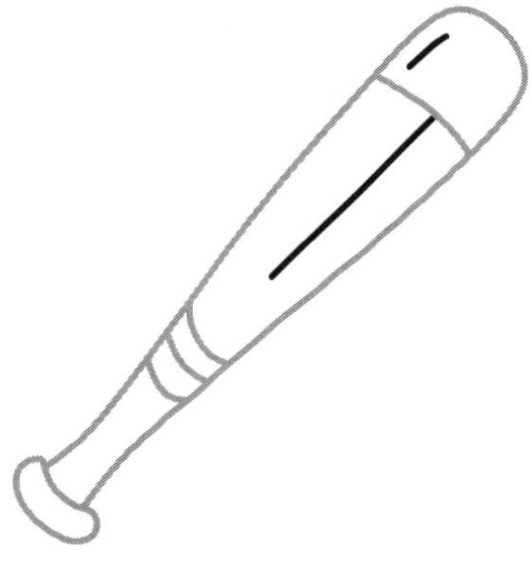

Now, try to draw

Jersey
Jerseys with numbers show who each player is on the field or court
1
2
3
4
4
5
4
Now, try to draw

Sneakers

Sports sneakers help athletes run faster and jump higher

1

2

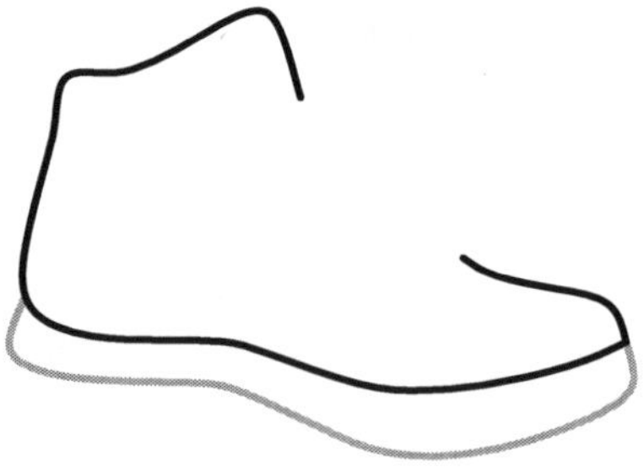

3

4

5

Now, try to draw

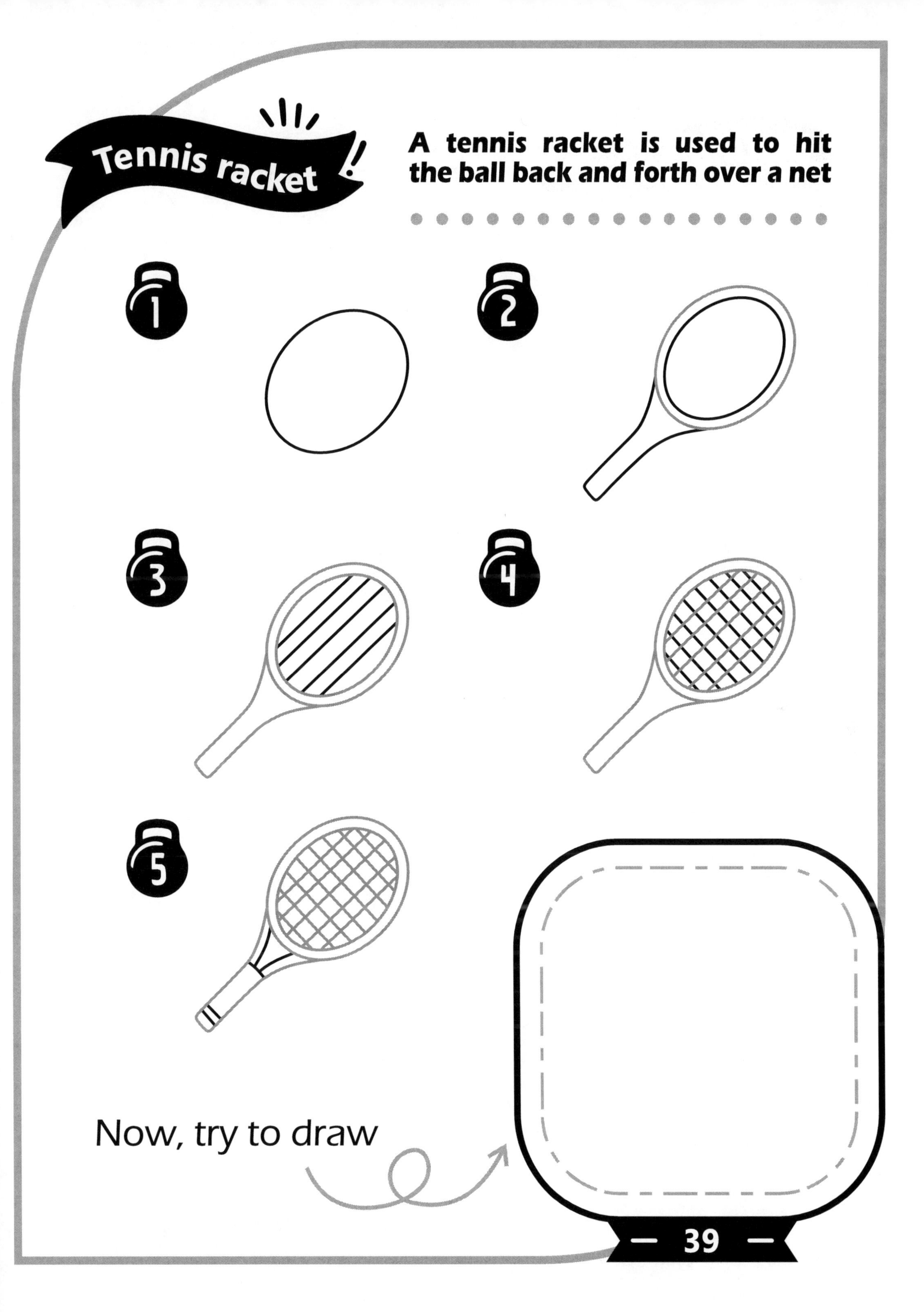
Tennis racket
A tennis racket is used to hit the ball back and forth over a net
1
2
3
4
5
Now, try to draw

Boxing gloves!

Boxing gloves protect a boxer's hands during a match

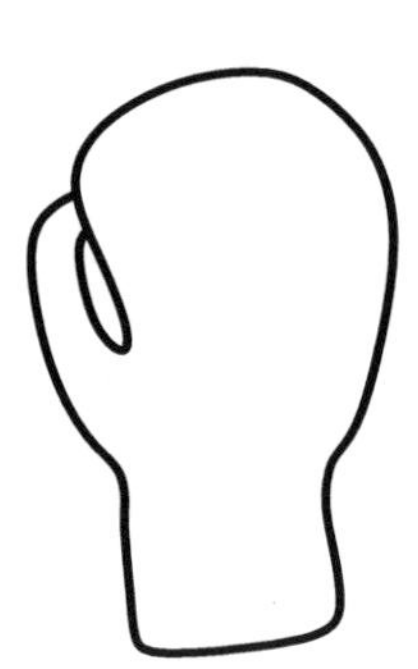

2

3

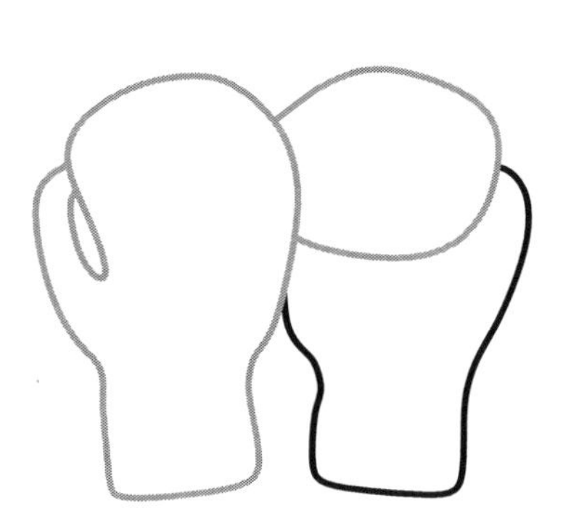

4

5

Now, try to draw

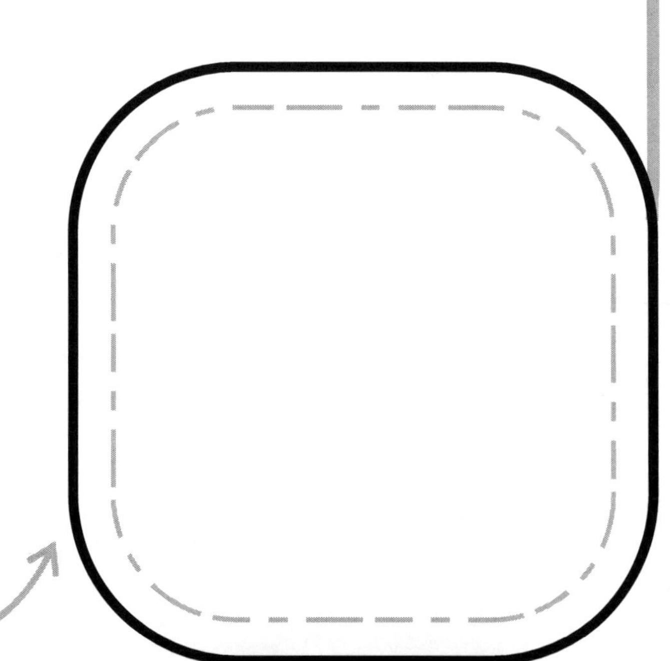

A rugby ball is oval-shaped and is used in the fast-paced game of rugby

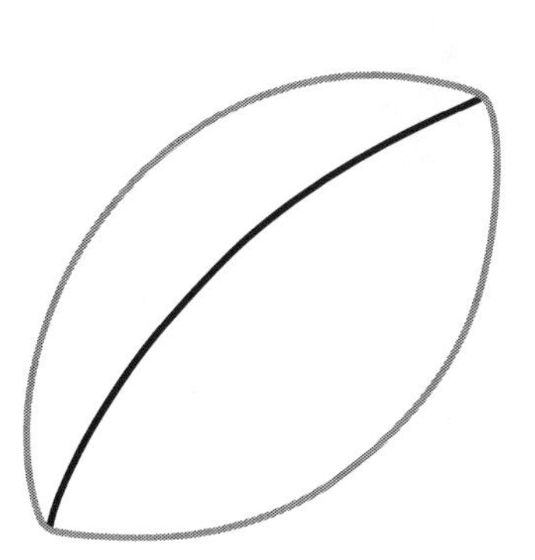

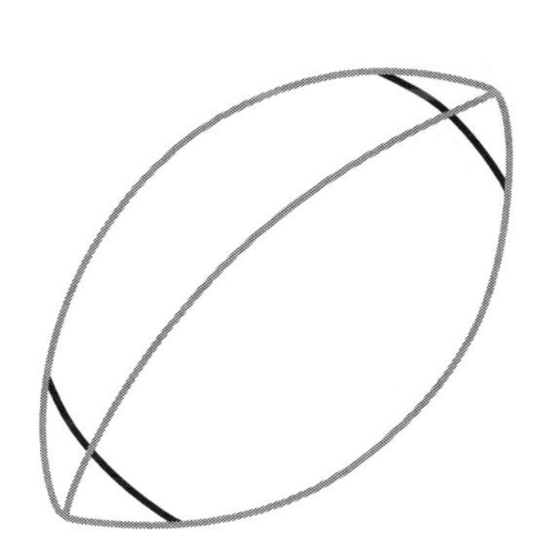

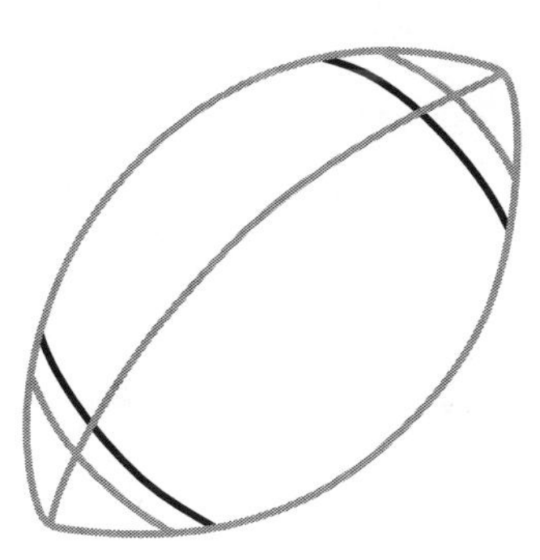

Now, try to draw

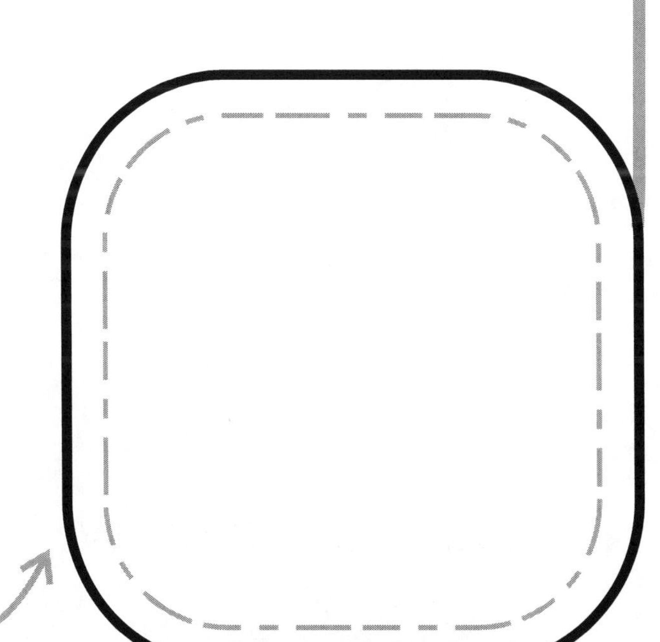

Football goals are the target where players aim to score points

1

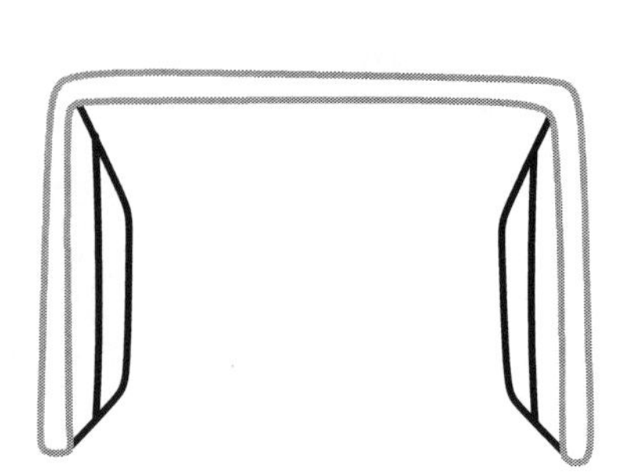

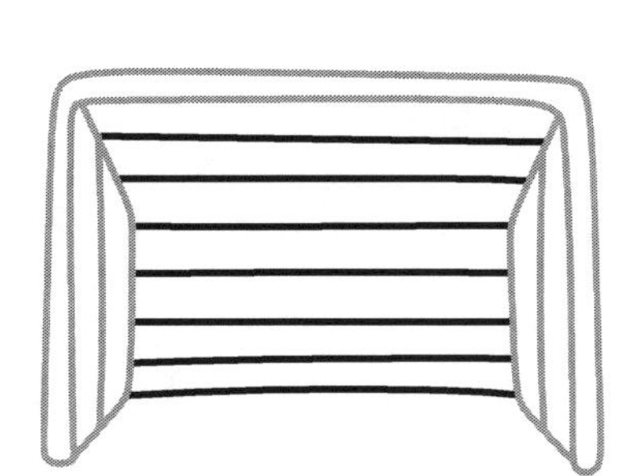

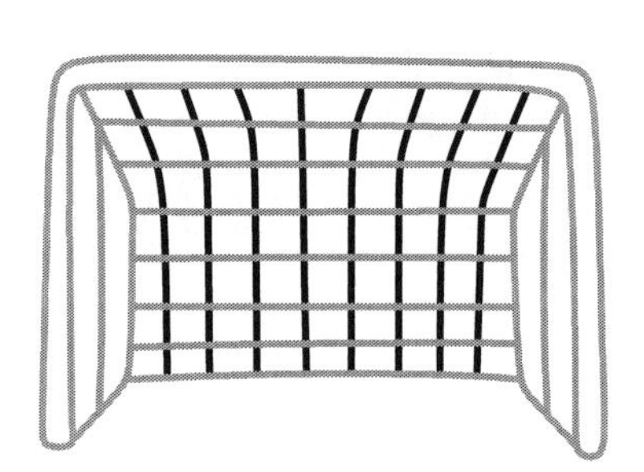

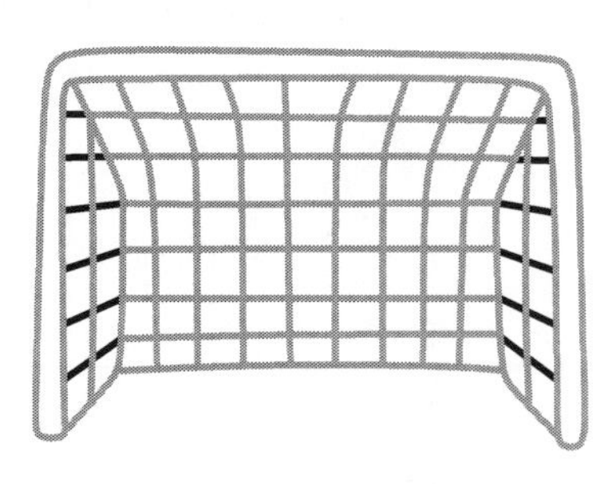

Now, try to draw

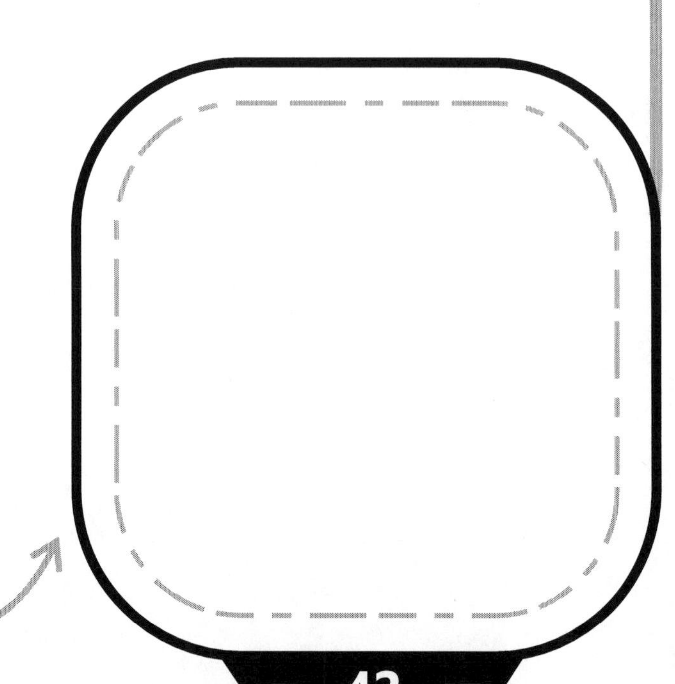

Cap
Caps are often worn by players to keep the sun out of their eyes during outdoor games
1
2
3
4
5
Now, try to draw

Foosball table

A foosball table is used for a fun game where players control little figures to score goals

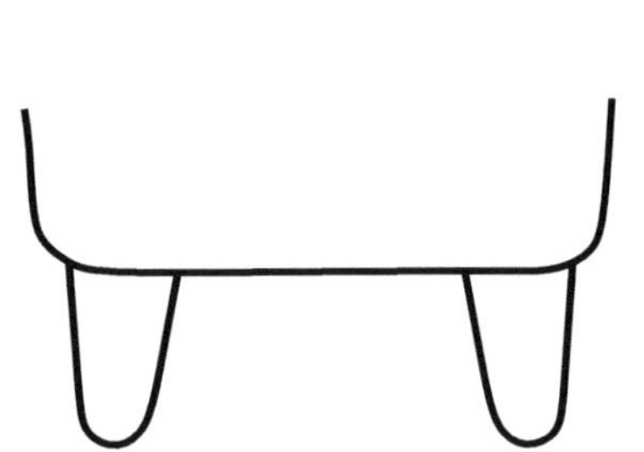

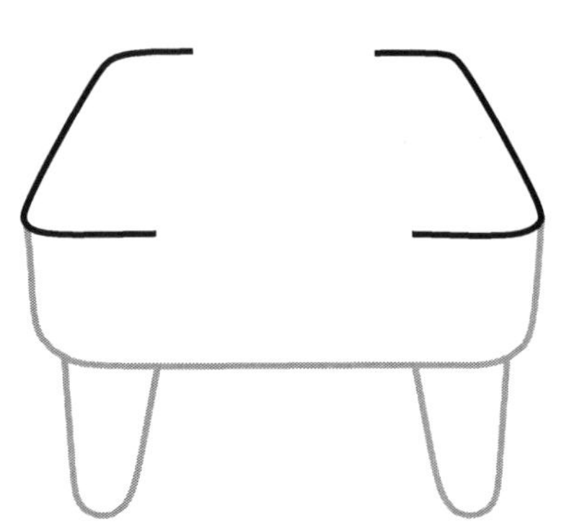

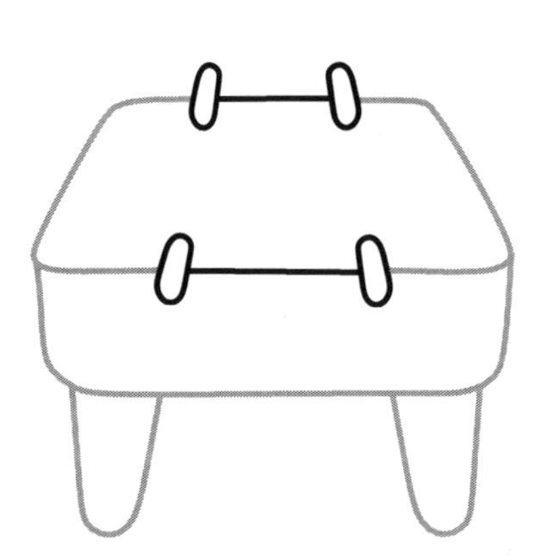

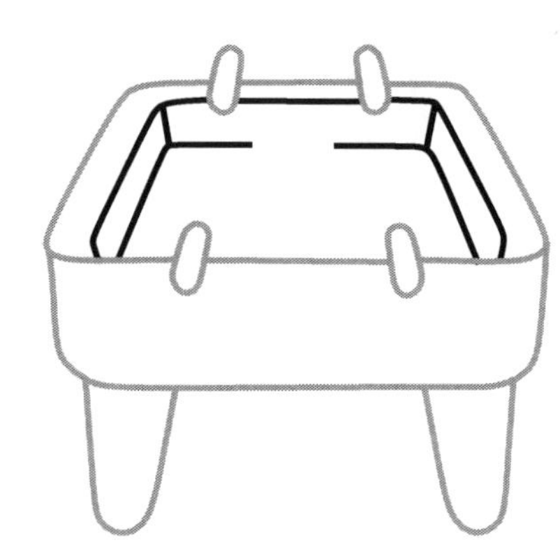

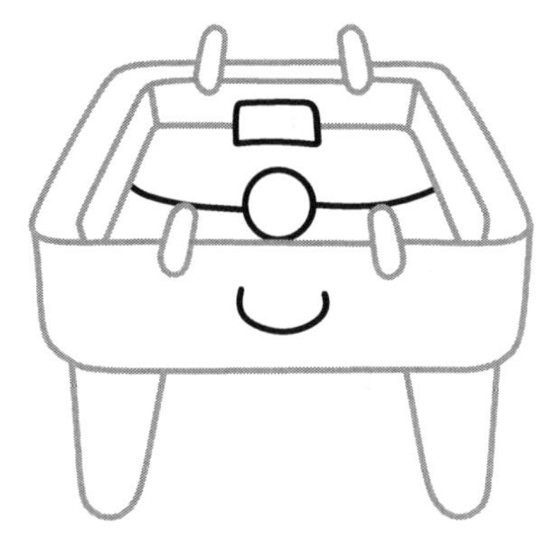

Now, try to draw

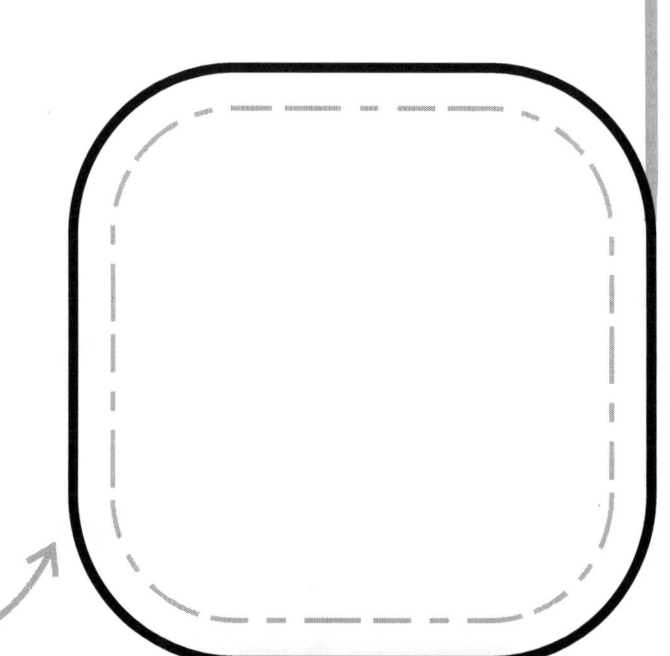

The championship cup is the ultimate prize for winning a big sports tournament!

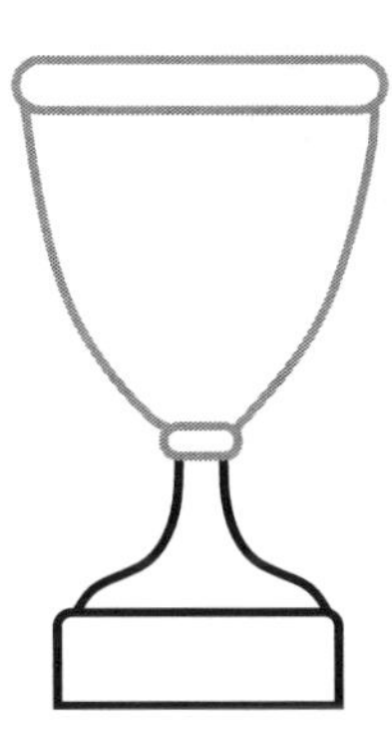

Now, try to draw

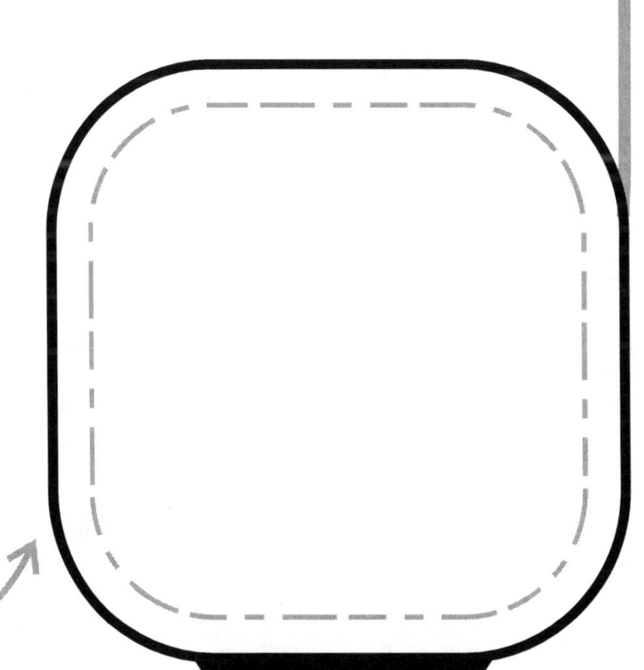

Sports medal!

Medals are awarded to the best athletes in competitions, especially at the Olympics

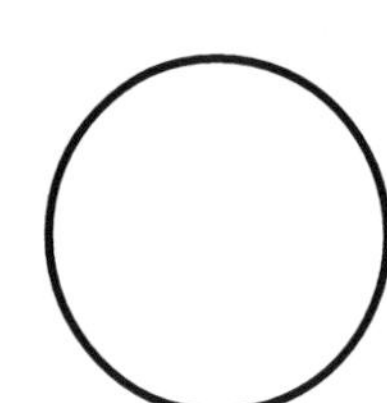

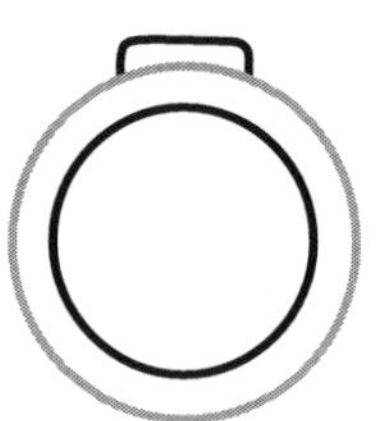

Now, try to draw

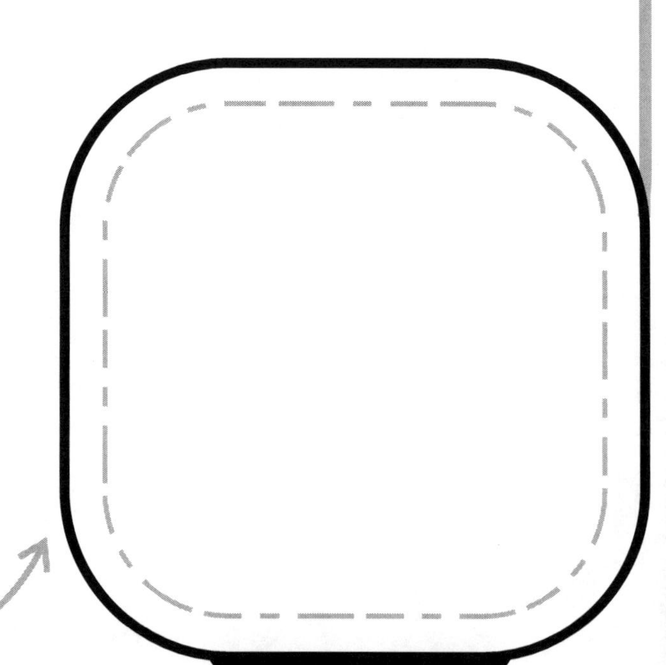

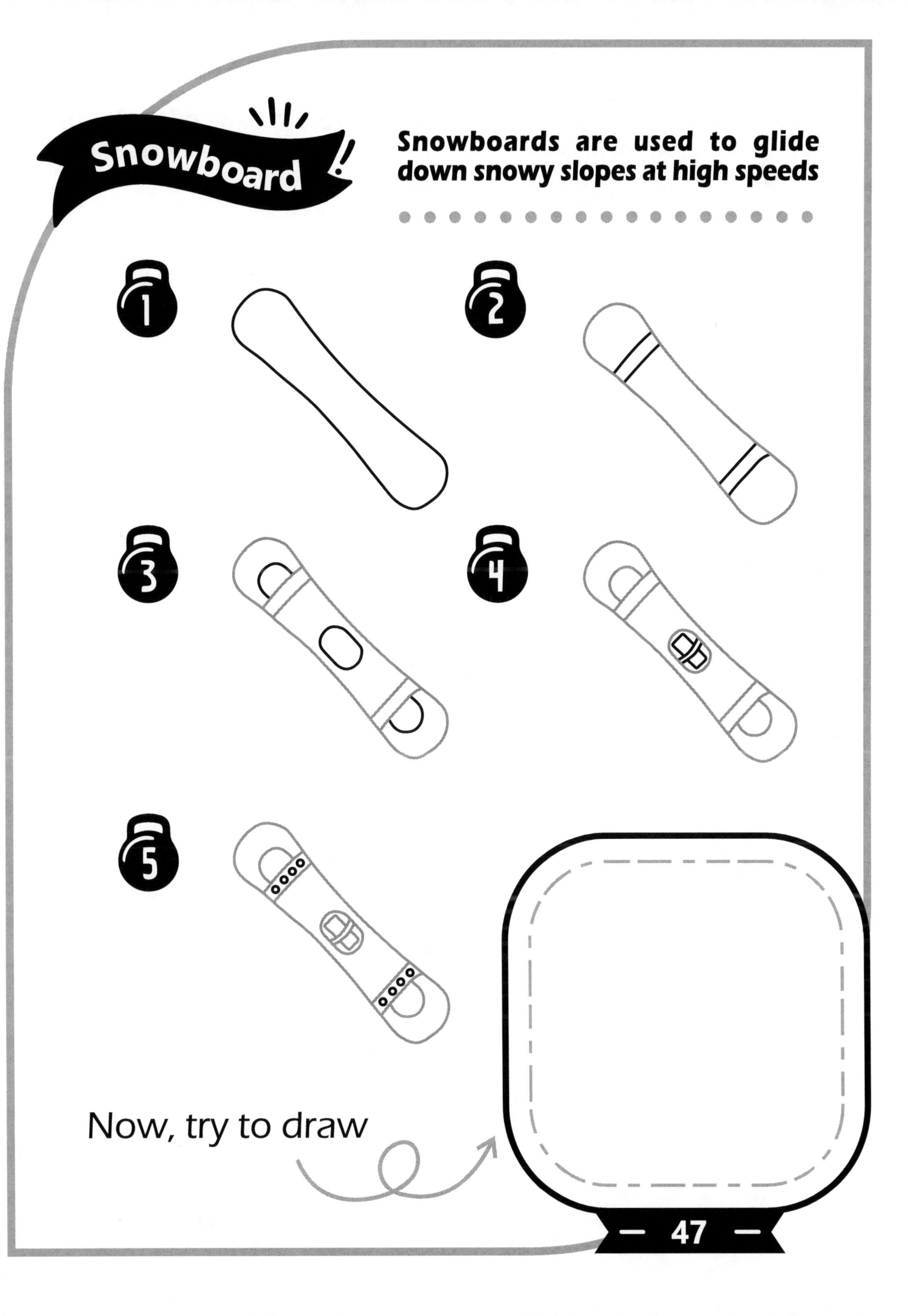
Snowboard
Snowboards are used to glide down snowy slopes at high speeds
1
2
3
4
5
Now, try to draw

Rollerblades have wheels in a line and are great for skating on sidewalks

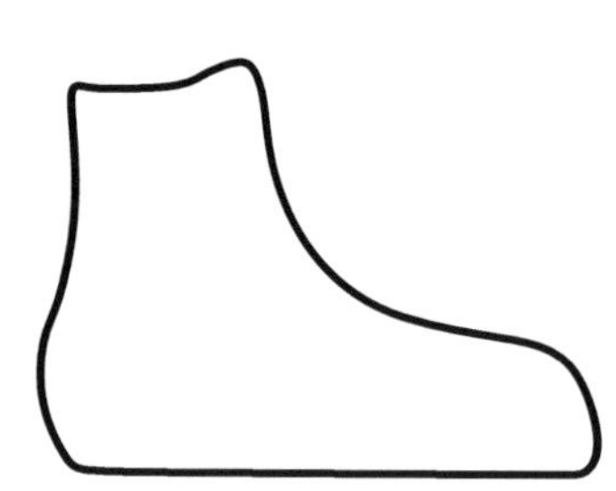

Now, try to draw

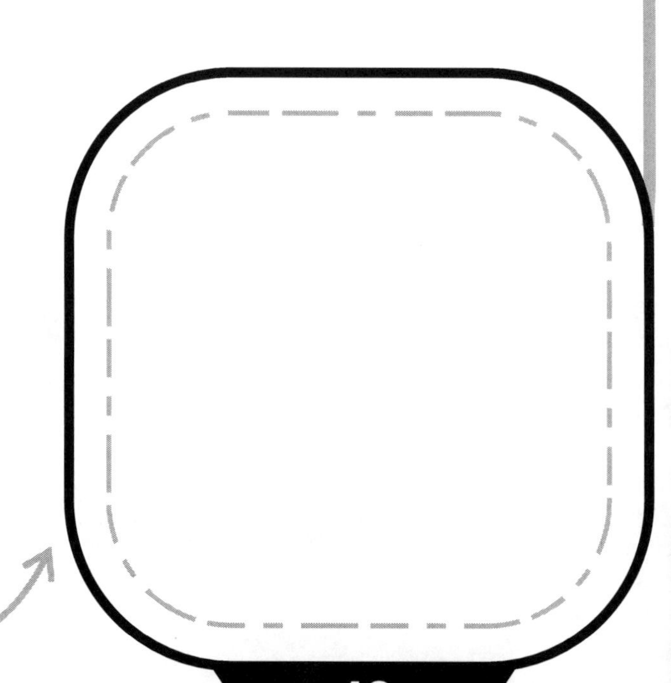

Frisbee
Frisbees are flying discs that are fun to throw and catch with friends
1
2
3
4
5
Now, try to draw

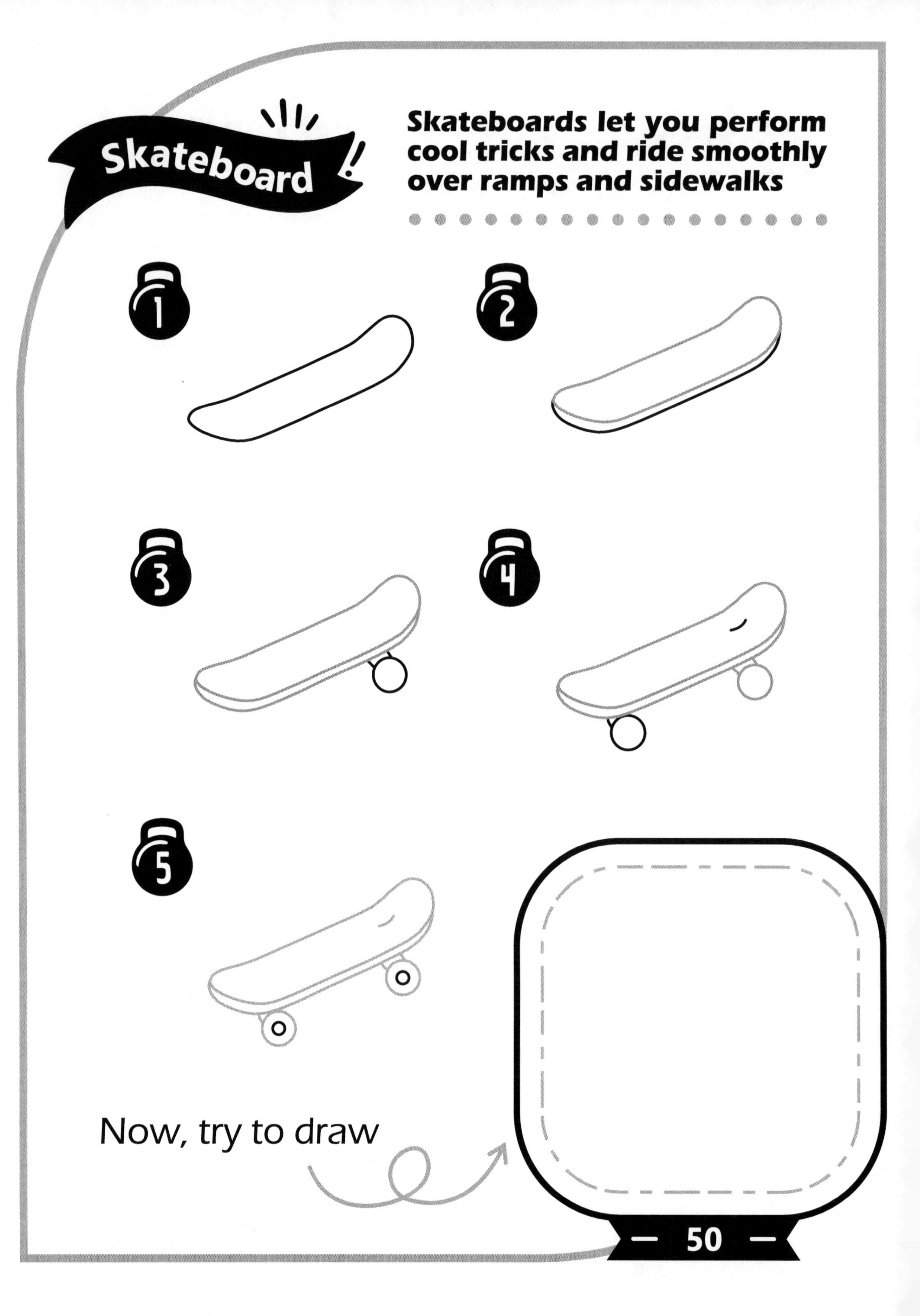
Skateboard
Skateboards let you perform cool tricks and ride smoothly over ramps and sidewalks
1
2
3
4
5
Now, try to draw

Superheroes with capes often fly or save the day with their amazing powers!

Now, try to draw

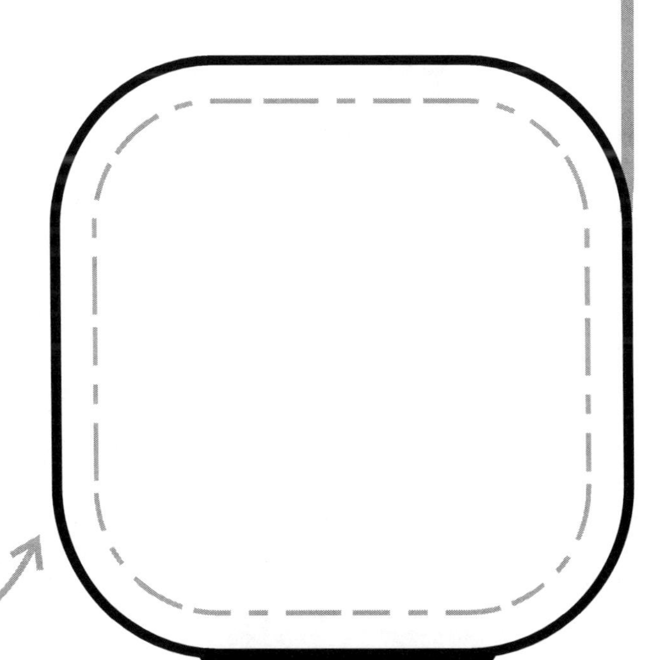

Knights wore armor to protect themselves during battles in the Middle Ages

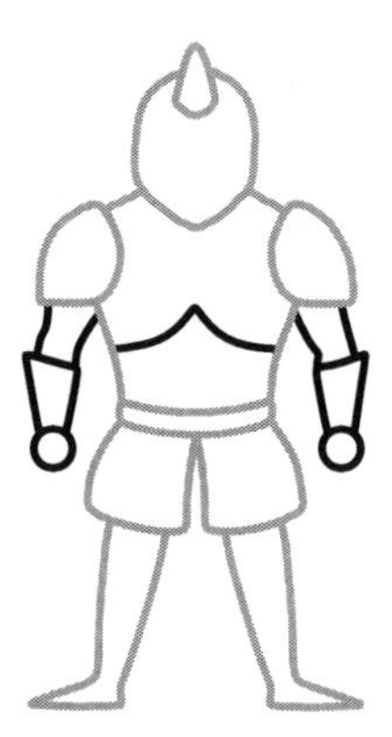

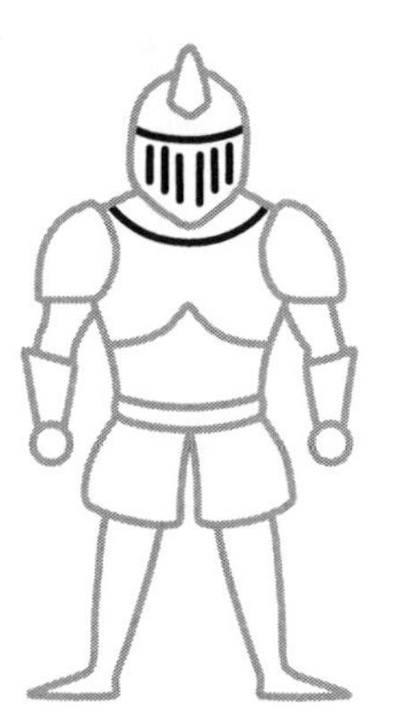

Now, try to draw

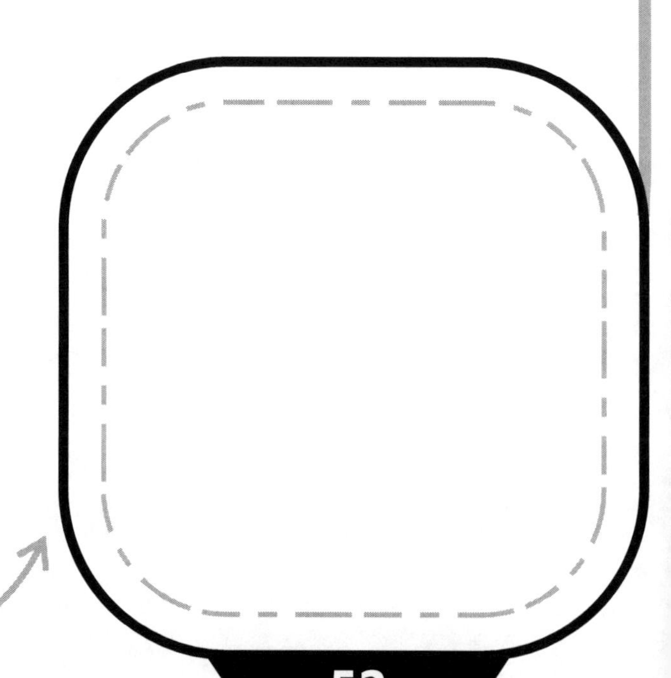

Ninjas are stealthy warriors skilled in martial arts and secret missions

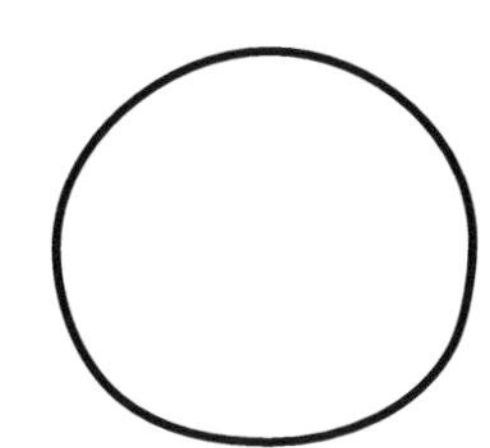

Now, try to draw

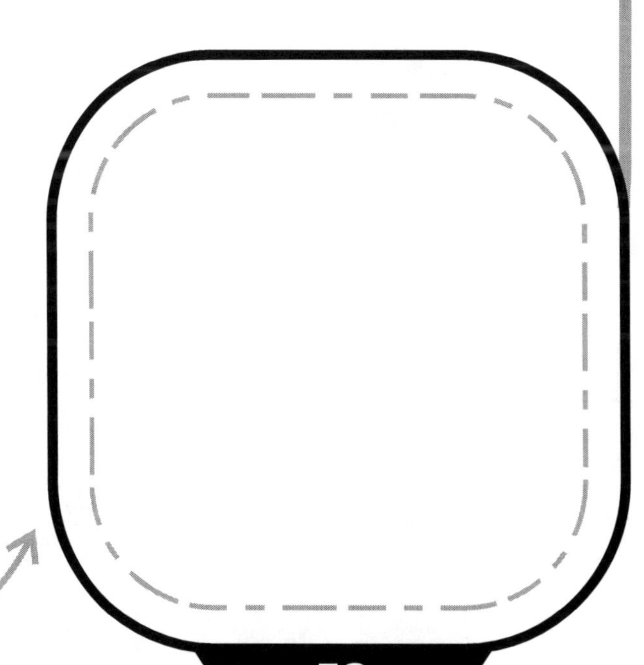

Wizards use magic spells to cast enchantments and perform incredible feats

Now, try to draw

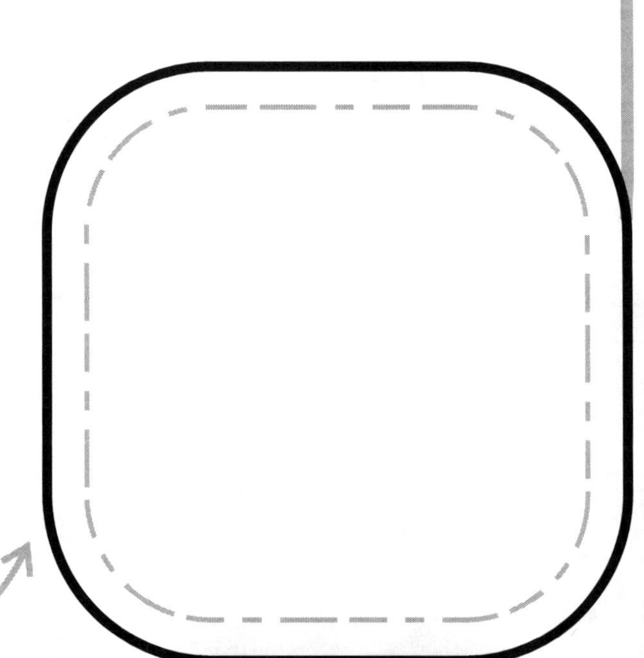

Vikings were fierce warriors and sailors who explored distant lands

Now, try to draw

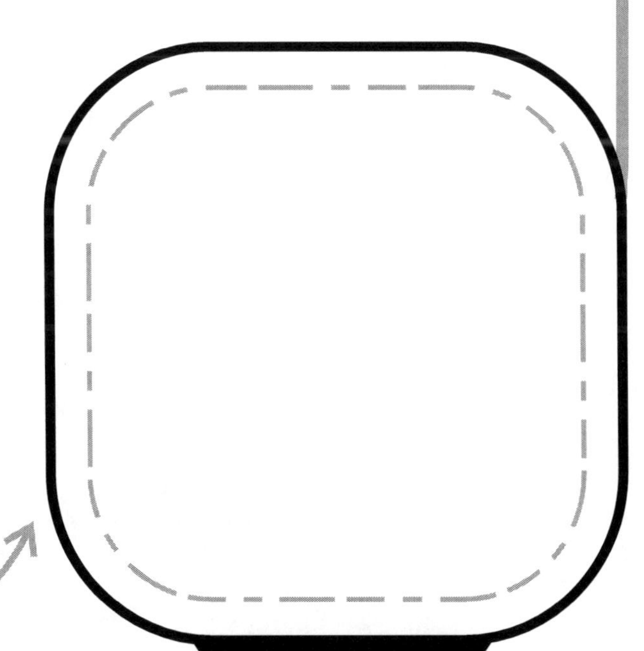

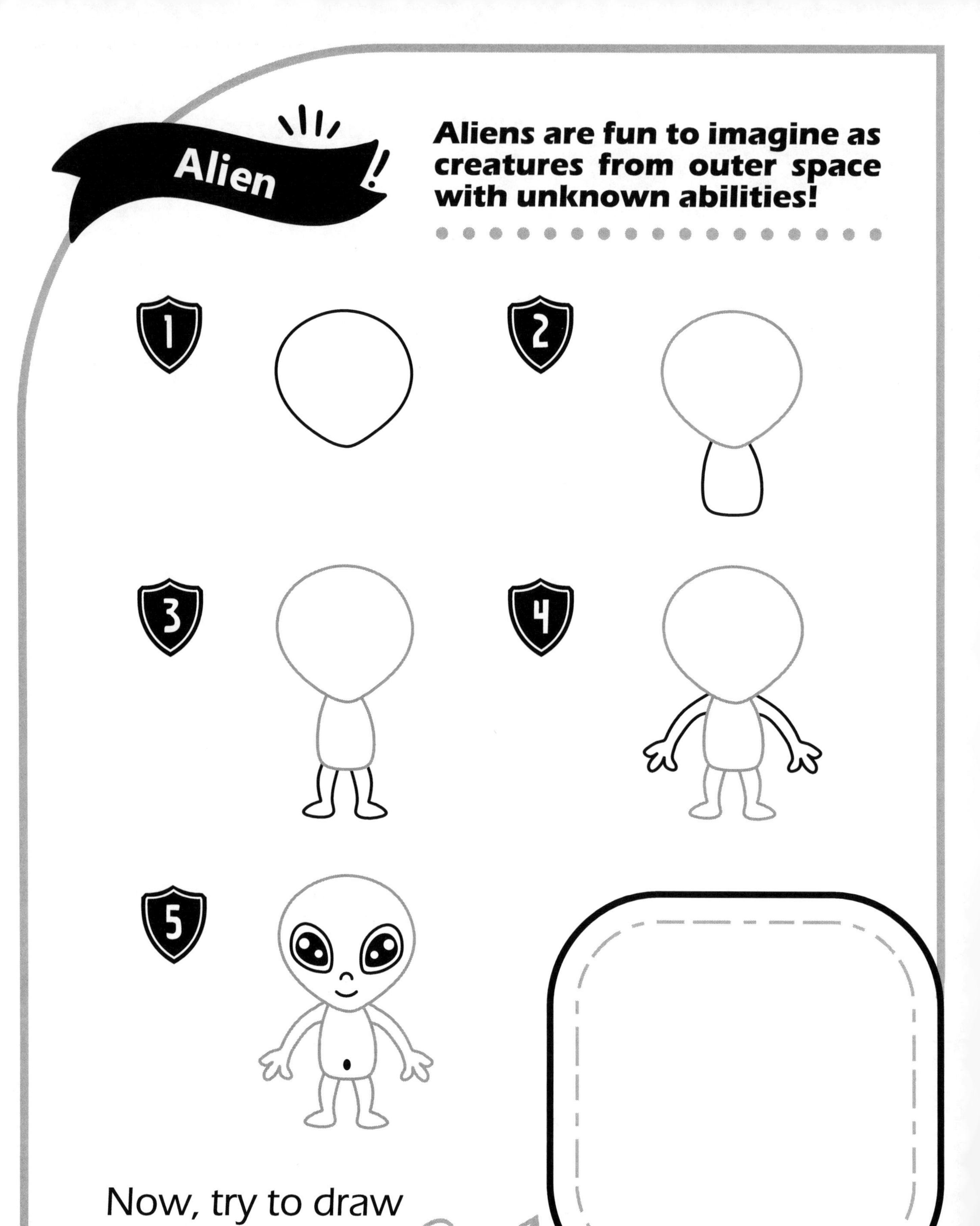
Alien
Aliens are fun to imagine as creatures from outer space with unknown abilities!
1
2
3
4
5
Now, try to draw

Dragons are legendary creatures that breathe fire and fly through the skies

Now, try to draw

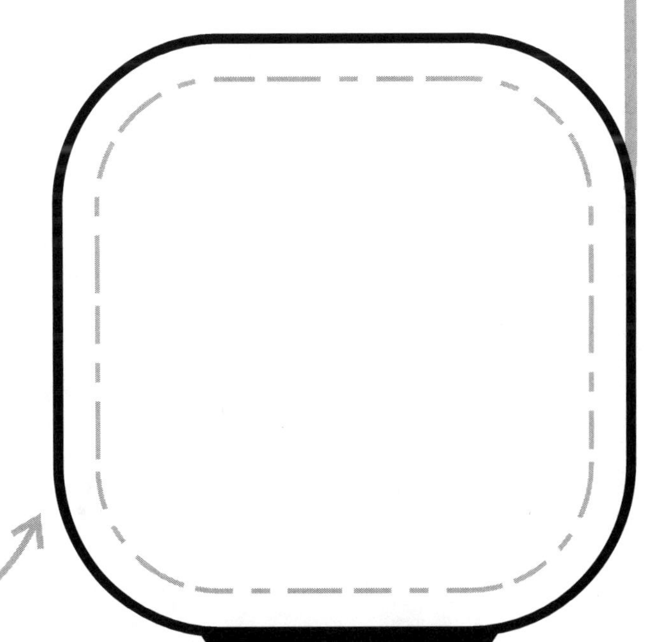

Transformers are robots that can change into cars, planes, and other vehicles!

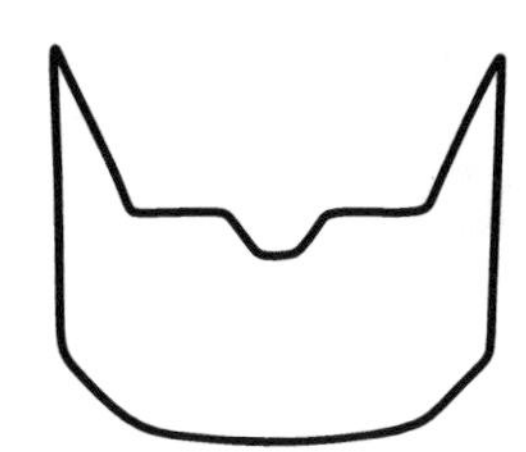

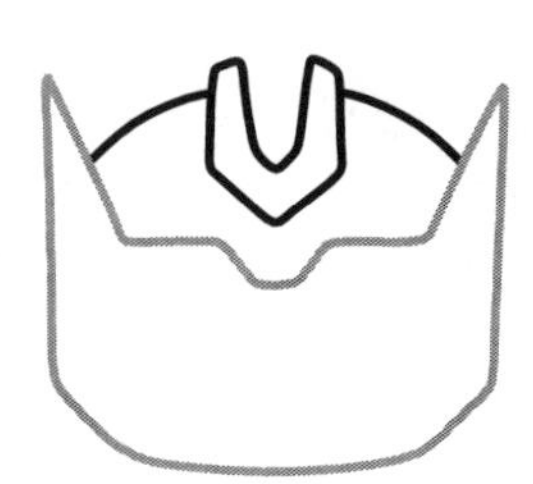

Now, try to draw

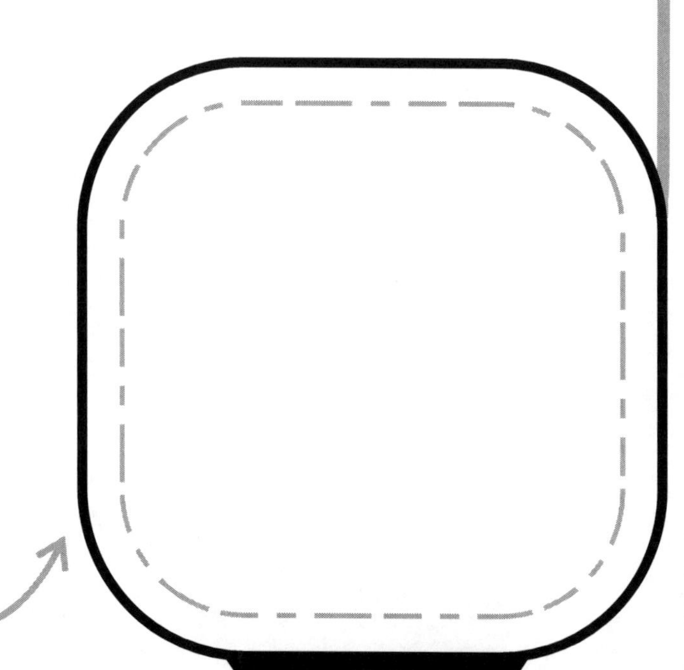

Drones can fly without a pilot inside and are often used for filming or racing

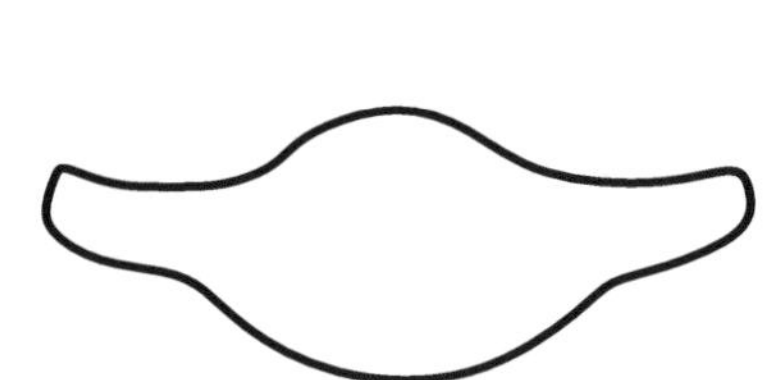

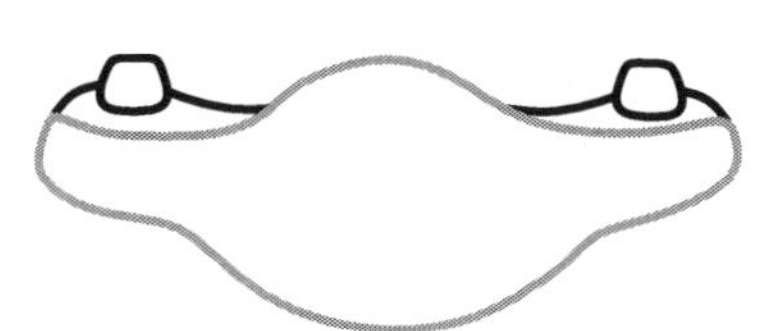

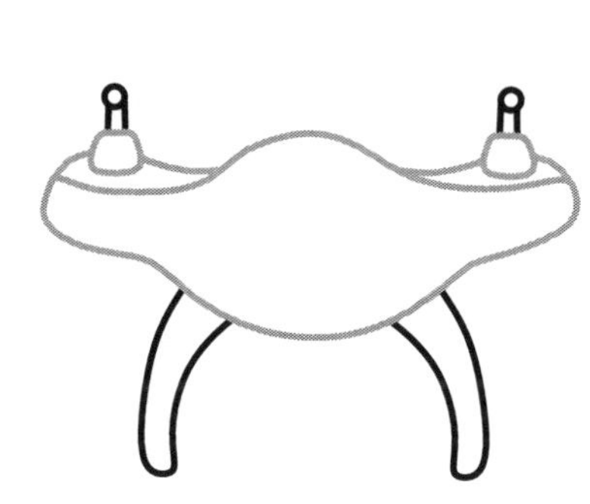

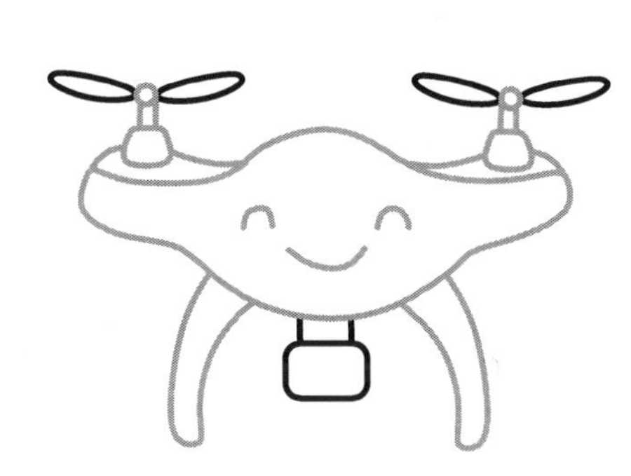

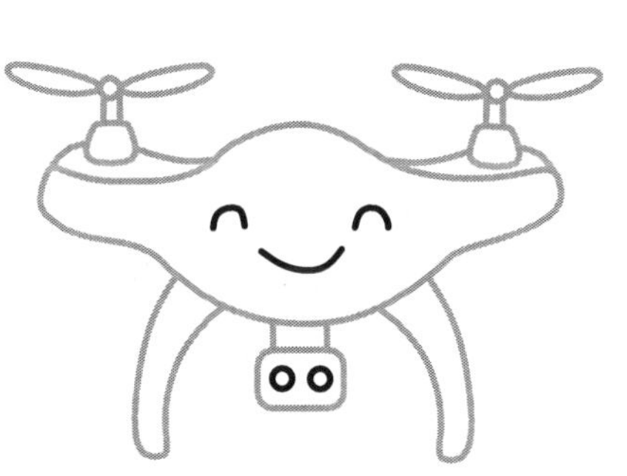

Now, try to draw

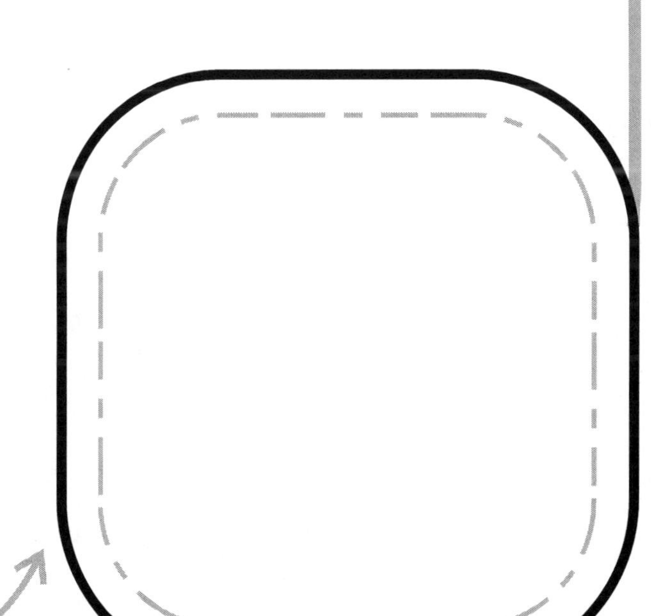

Robot dinosaur!

Robot dinosaurs combine prehistoric creatures with futuristic technology!

Now, try to draw

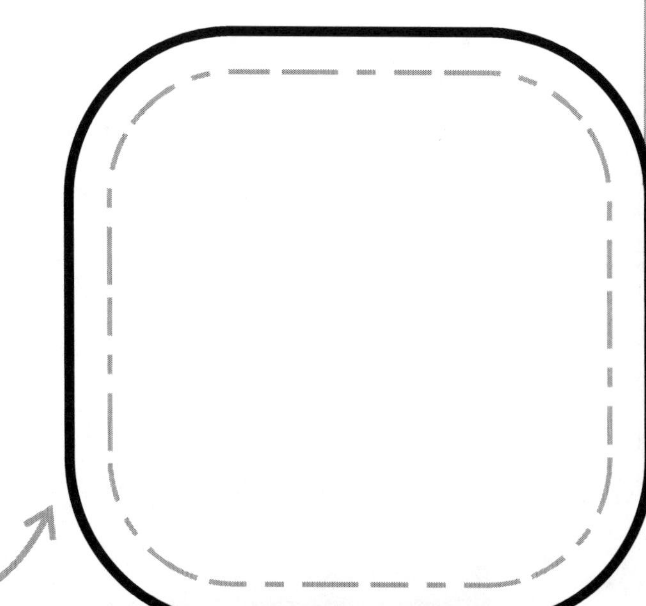

Griffins are mythical creatures with the body of a lion and the head and wings of an eagle

Now, try to draw

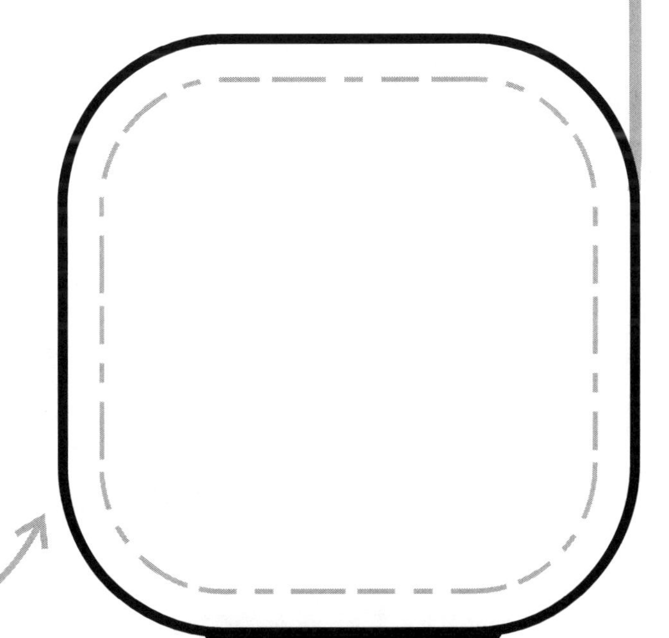

Pegasus is a magical horse with wings that can soar through the clouds

Now, try to draw

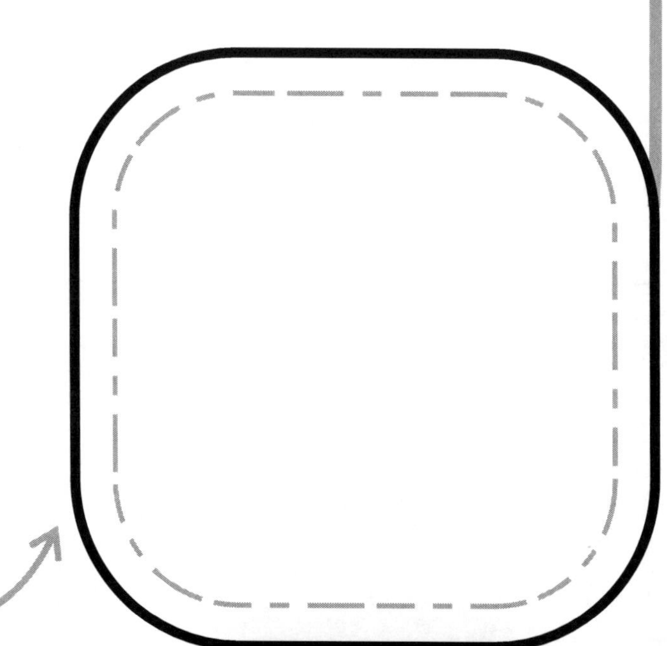

Giant spider!

Giant spiders often appear in stories and movies, known for their creepy, crawly webs

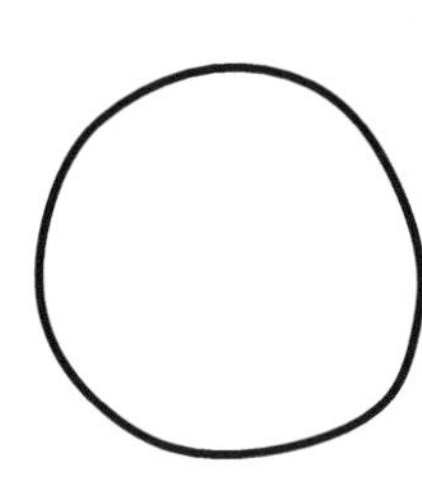

Now, try to draw

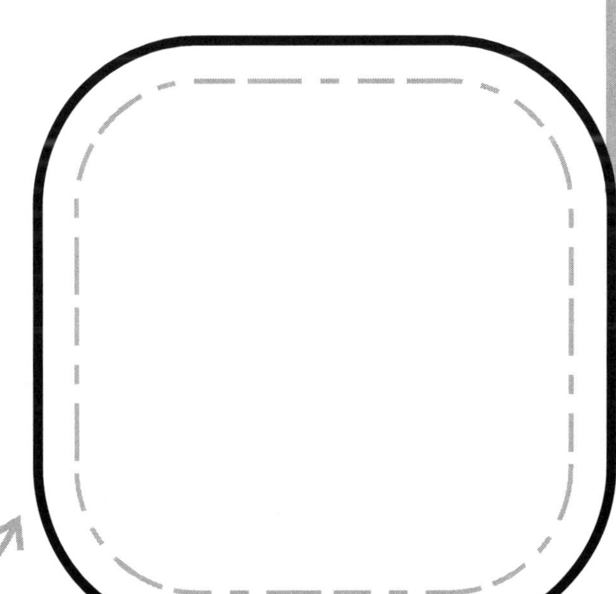

Centaurs are mythical beings with the upper body of a human and the lower body of a horse

Now, try to draw

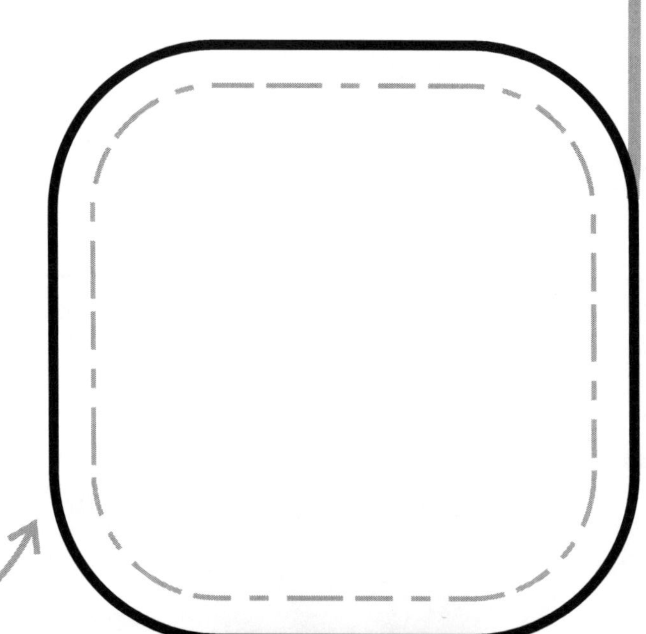

Pumpkin!

Pumpkins are often carved into spooky faces for Halloween!

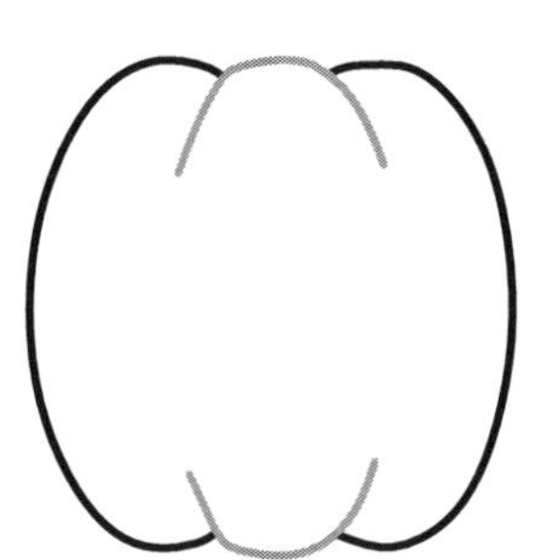

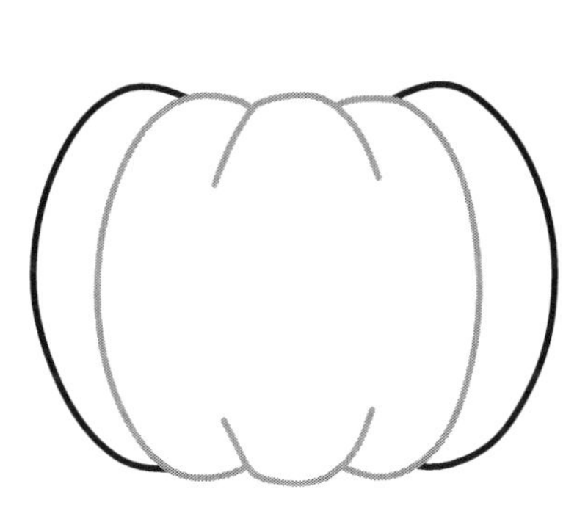

Now, try to draw

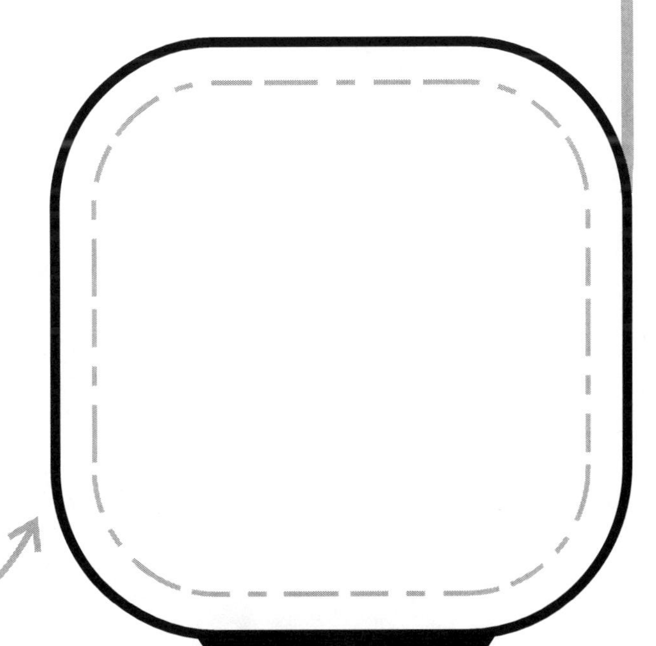

Ghost
Ghosts are mysterious figures in folklore, often appearing as floating, see-through shapes
1
2
3
4
5
Now, try to draw

Pirate

Pirates sailed the seas, searching for treasure and adventure

Now, try to draw

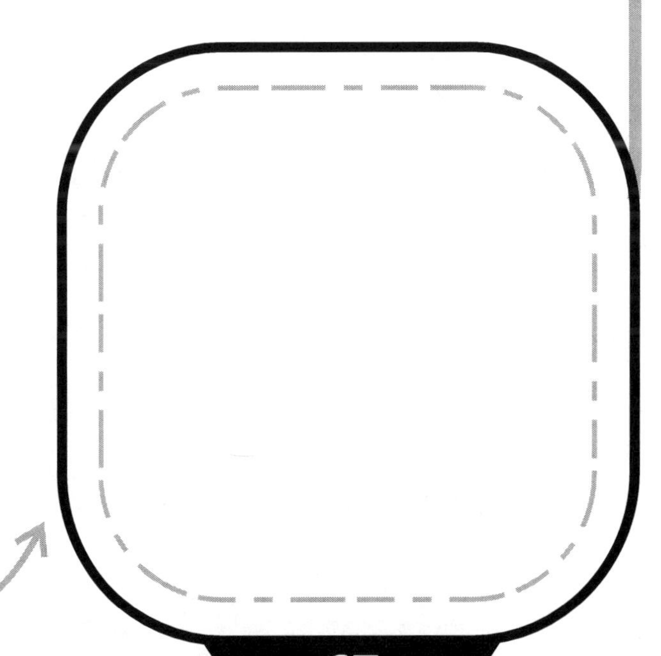

Tiger
Tigers are the largest wild cats in the world and have beautiful orange and black stripes
1
2
3
4
5
Now, try to draw

Wolves live and hunt in packs, working together to survive in the wild

1

2

3

4

5

Now, try to draw

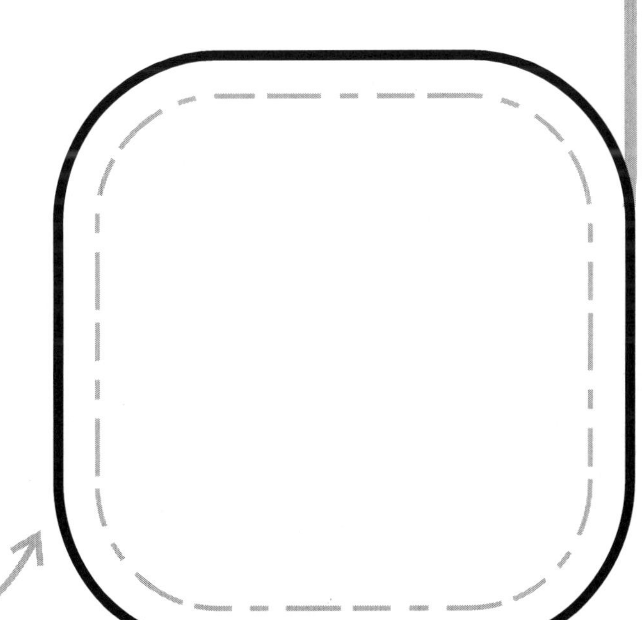

Elephants are the largest land animals, known for their long trunks and big ears

1

2

3

4

5

Now, try to draw

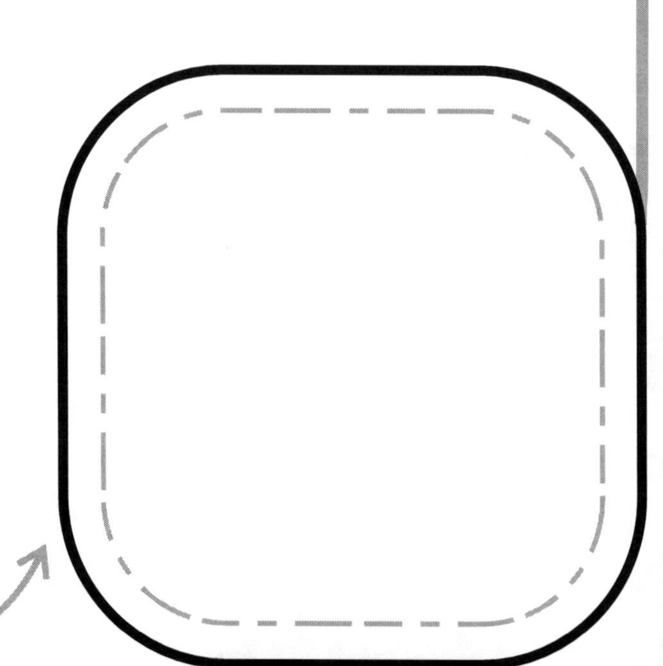

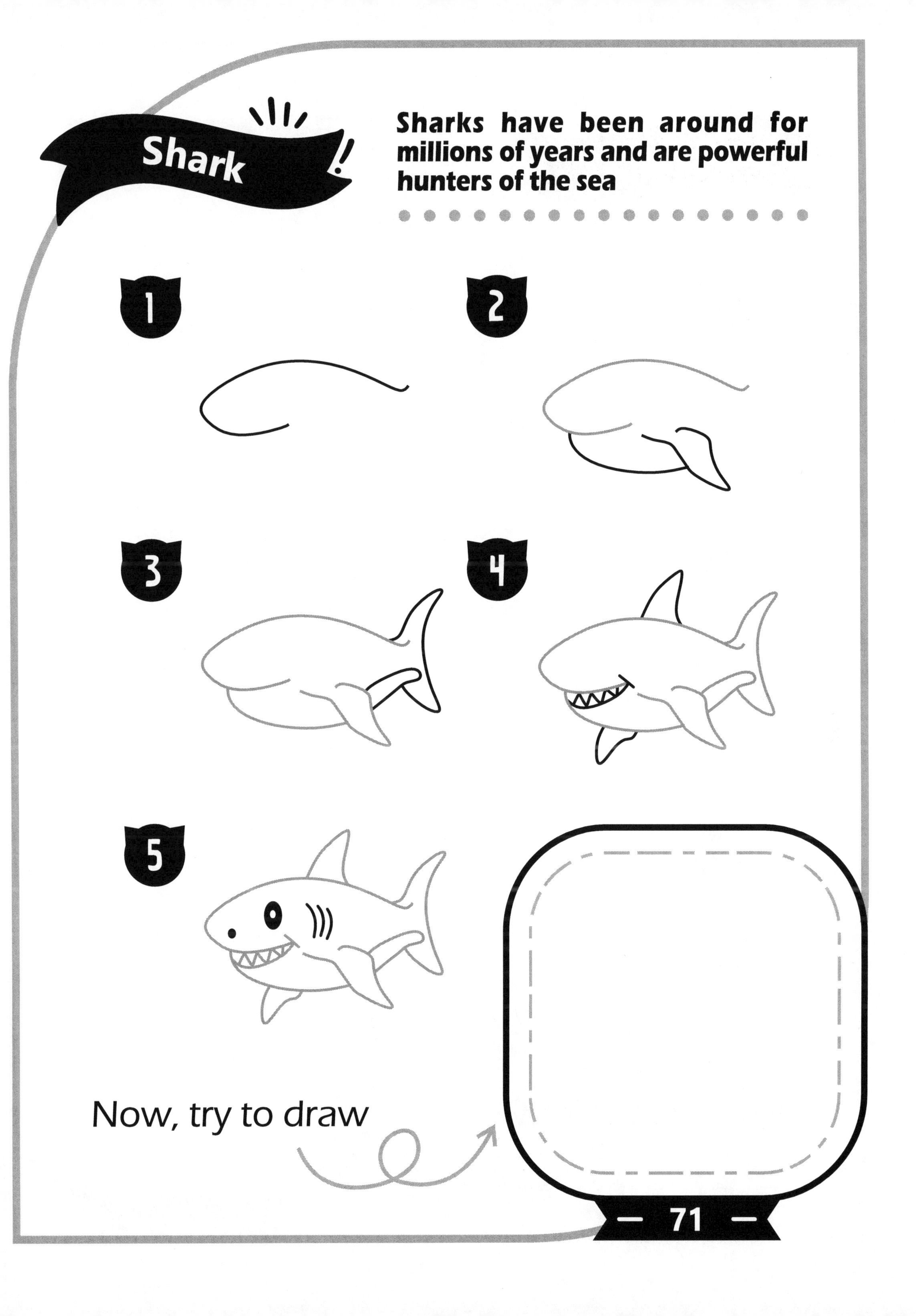
Shark
Sharks have been around for millions of years and are powerful hunters of the sea
1
2
3
4
5
Now, try to draw

Bear
Bears are strong animals that can stand on two legs and love eating fish and berries
1
2
3
4
5
Now, try to draw

Crocodiles have powerful jaws and can live both in water and on land

2

3

4

5

Now, try to draw

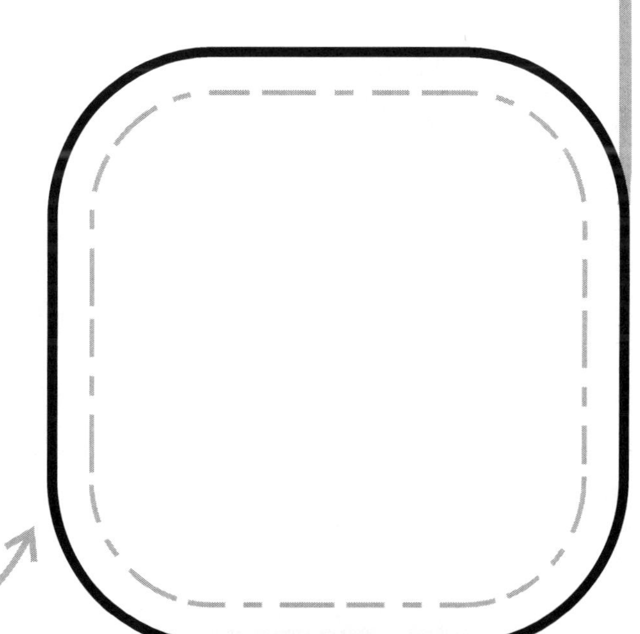

Eagles have amazing eyesight and can spot prey from high in the sky

1

2

3

4

5

Now, try to draw

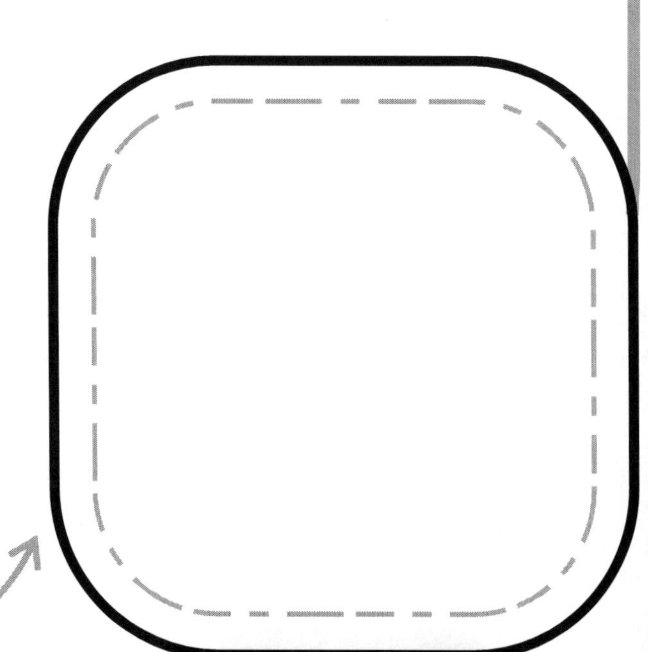

Dogs are loyal companions and often called "man's best friend

1

2

3

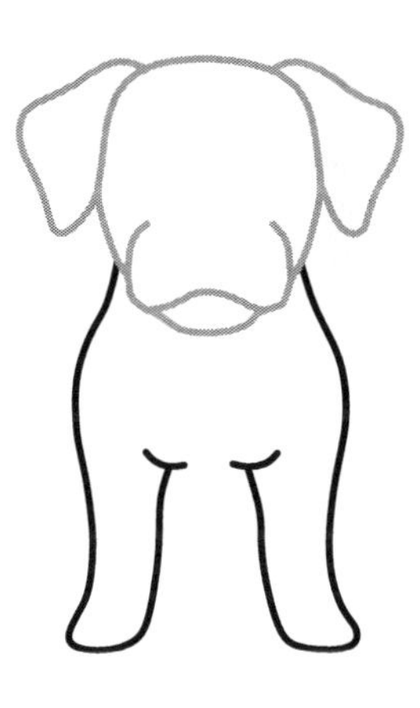

4

5

Now, try to draw

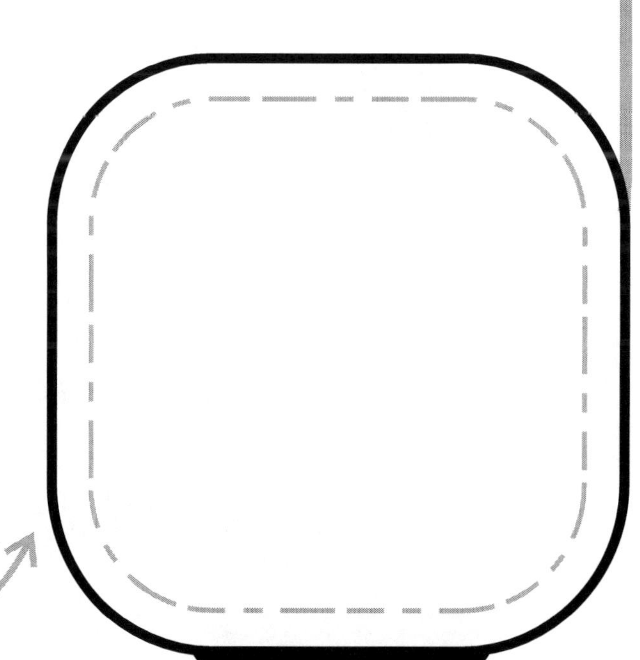

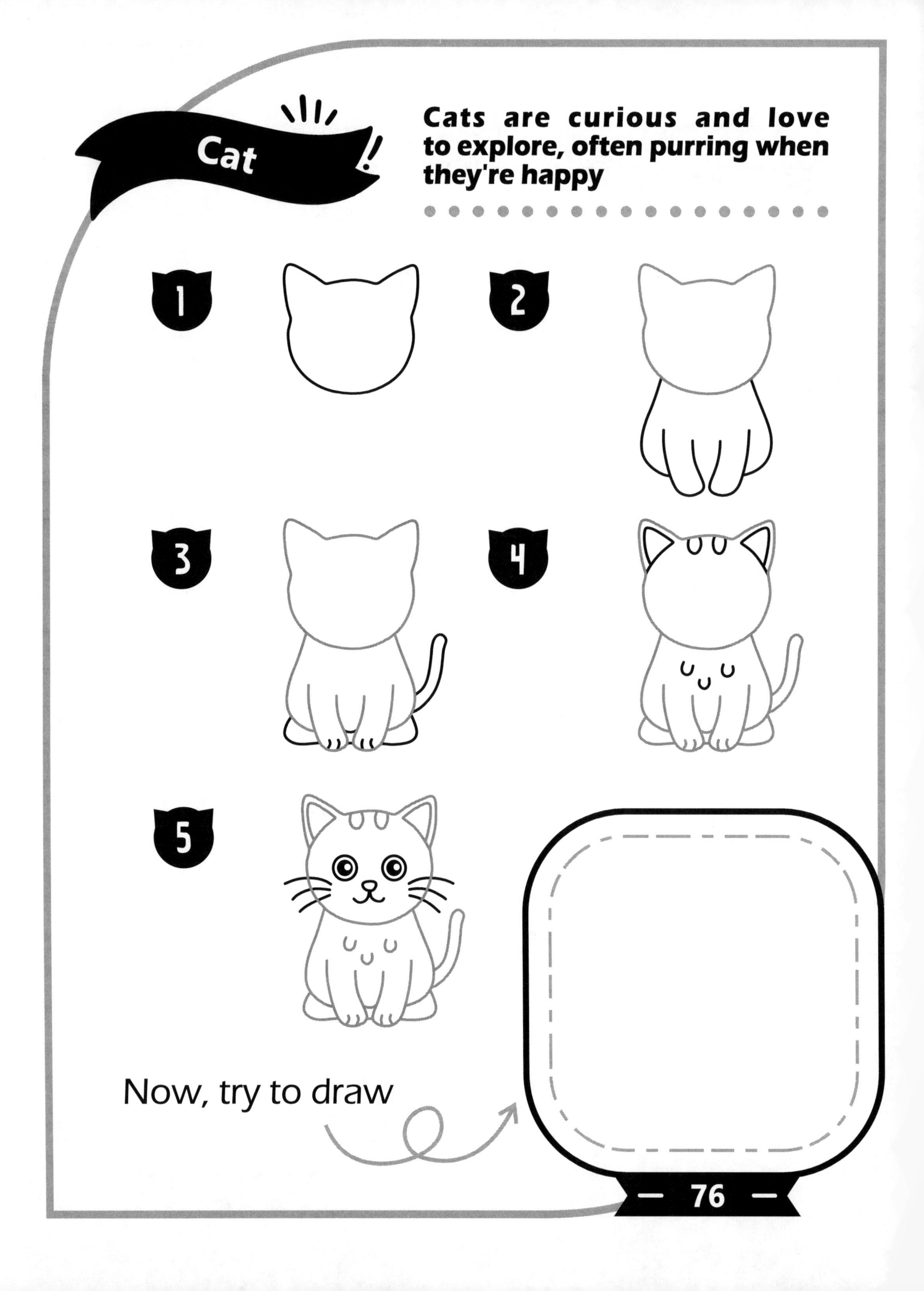
Cat
Cats are curious and love to explore, often purring when they're happy
1
2
3
4
5
Now, try to draw

Parrot
Parrots can mimic sounds and even talk, making them fun and colorful pets
1
2
3
4
5
Now, try to draw

Penguin

Penguins are flightless birds that waddle and swim in icy waters

1

2

3

4

5

Now, try to draw

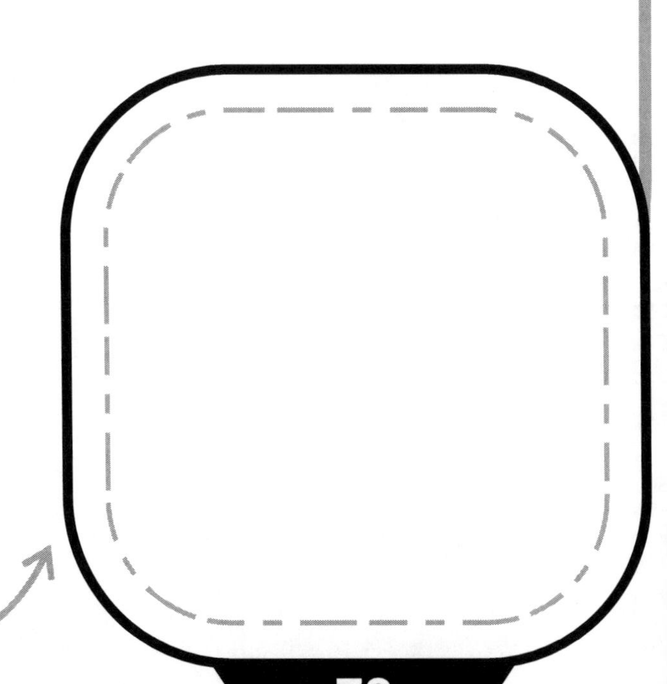

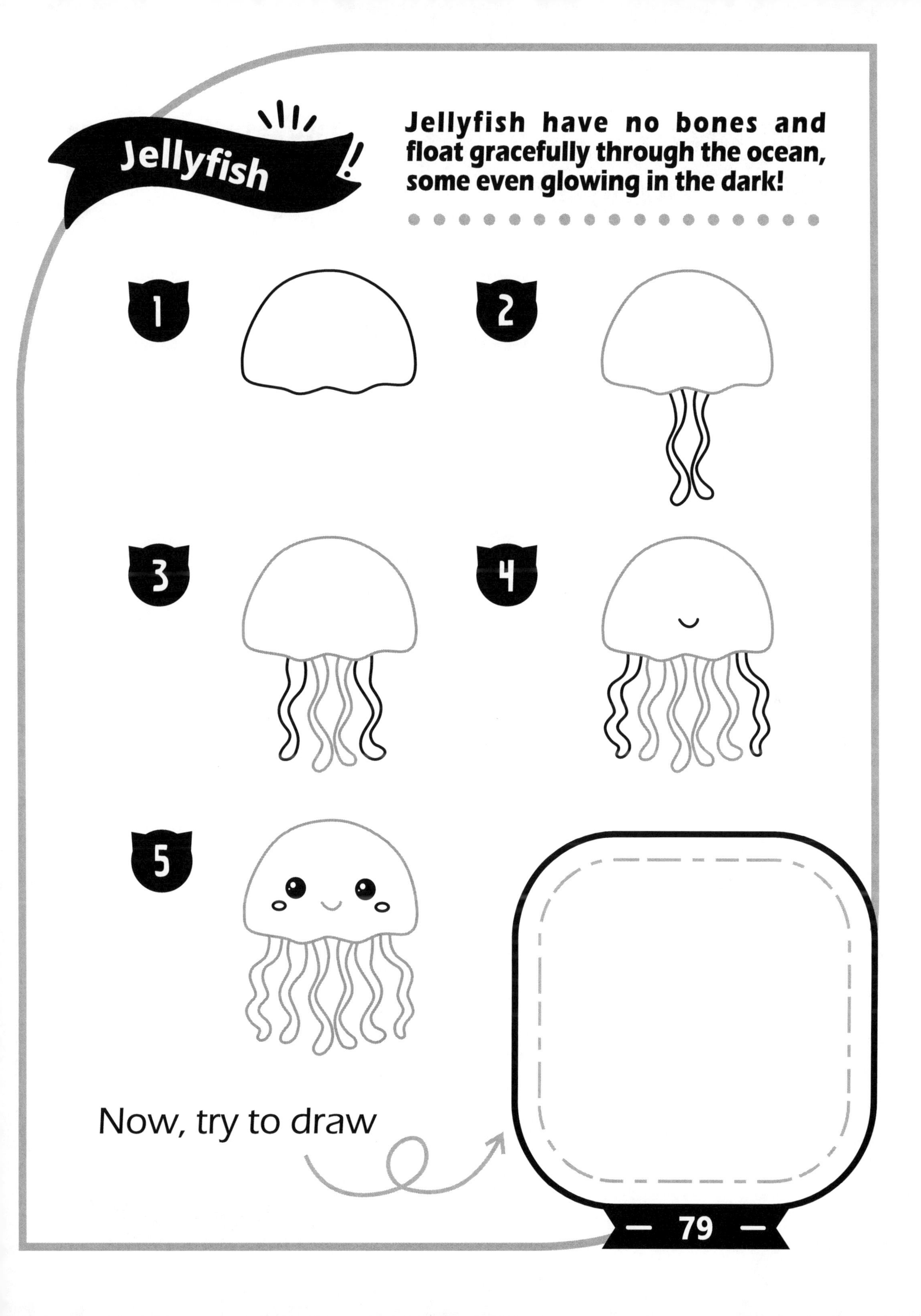
Jellyfish
Jellyfish have no bones and float gracefully through the ocean, some even glowing in the dark!
1
2
3
4
5
Now, try to draw

Sparkle the Little Fox!

Sparkle the Little Fox is a magical fox who loves adventures in enchanted forests!

1

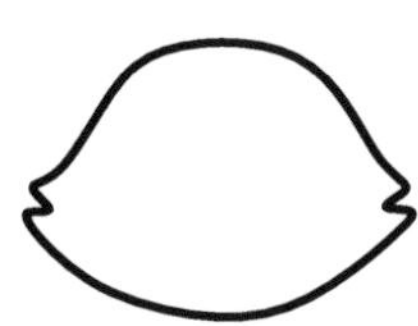

2

3

4

5

Now, try to draw

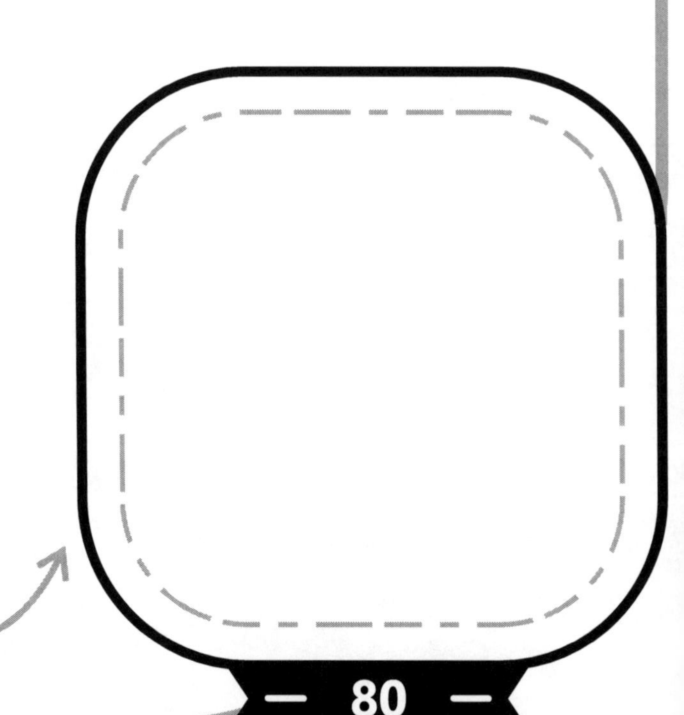

Lion
Lions are known as the "king the jungle" and live in family groups called prides
1
2
3
4
5
Now, try to draw

Gaming consoles let you play video games on your TV with controllers

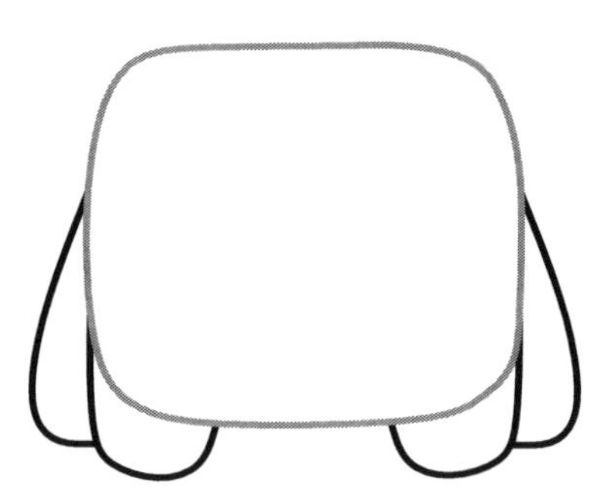

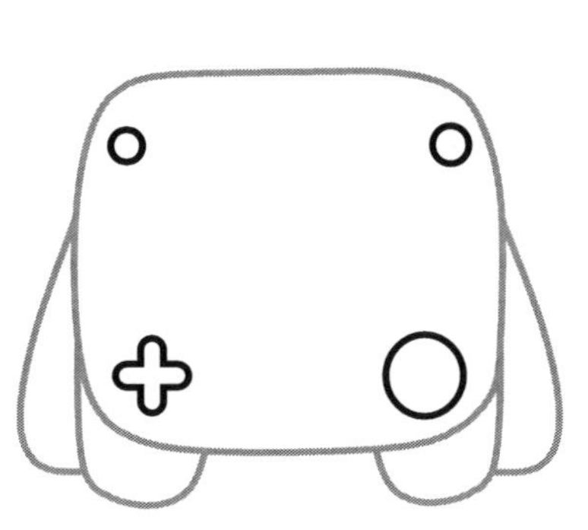

Now, try to draw

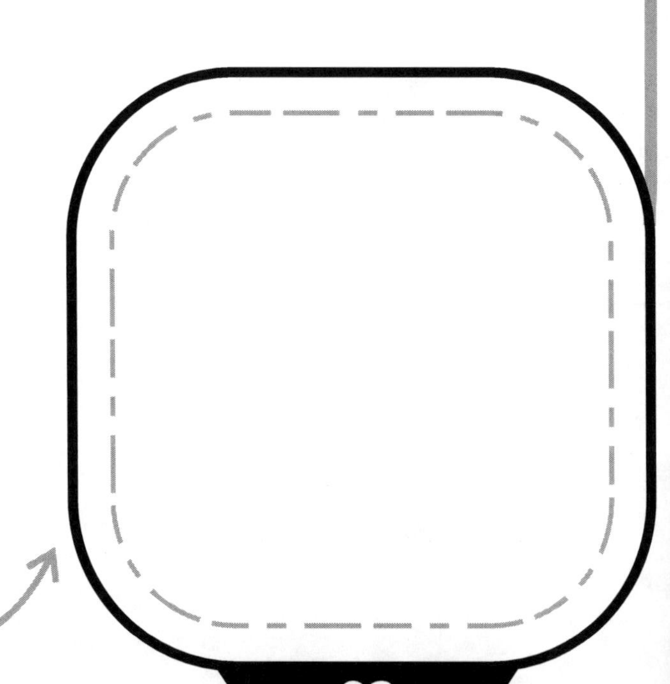

Lightsaber

A lightsaber is a glowing energy sword from the world of science fiction

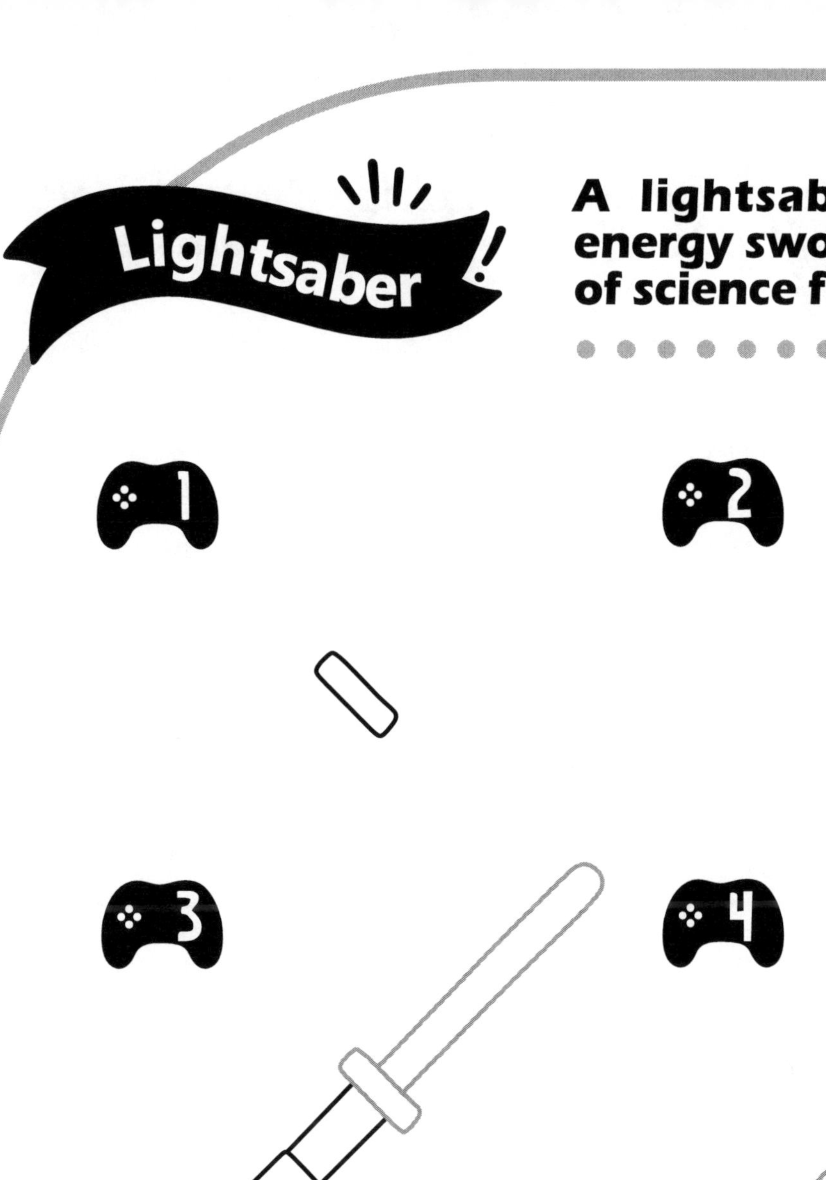

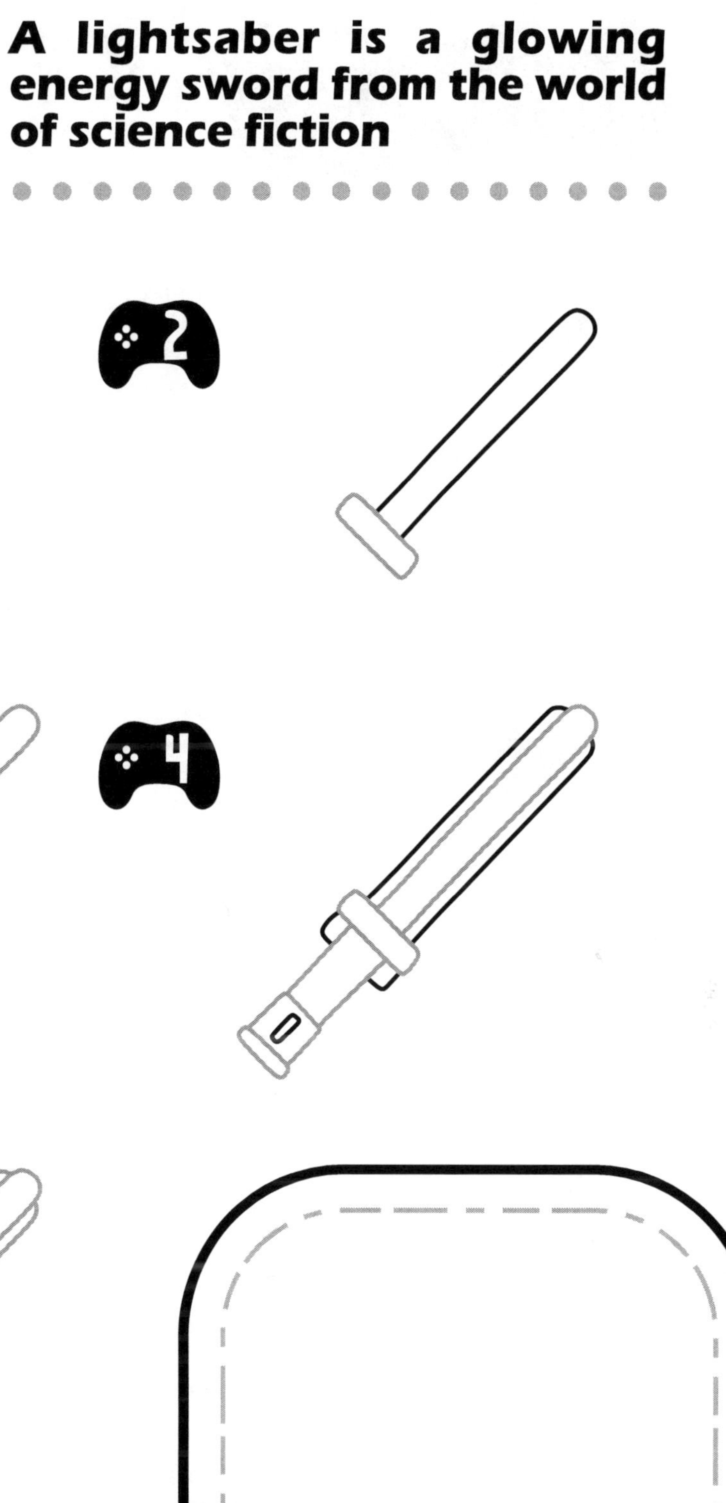

Now, try to draw

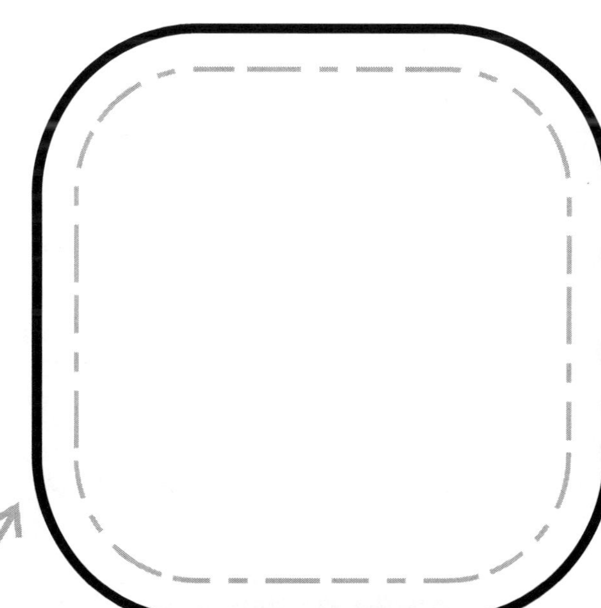

Smartphone!

Smartphones can do it all—call, text, take pictures, and even play games!

1

2

3

4

5

Now, try to draw

Laptop

Laptops are portable computers you can take anywhere

1

2

3

4

5

Now, try to draw

VR glasses let you step into a 3D world for games and adventures

Now, try to draw

Robot vacuums clean the floor all by themselves!

Now, try to draw

Headphones

Headphones let you listen to music or games without bothering others

Now, try to draw

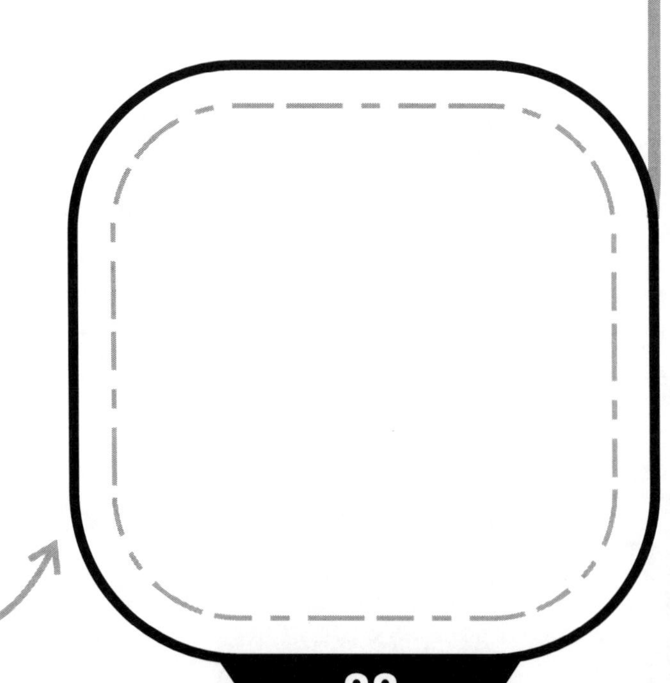

Cameras capture moments by taking photos and videos

Electric toothbrushes make brushing teeth easier and more fun!

Now, try to draw

Marker

Markers are colorful pens perfect for drawing or writing

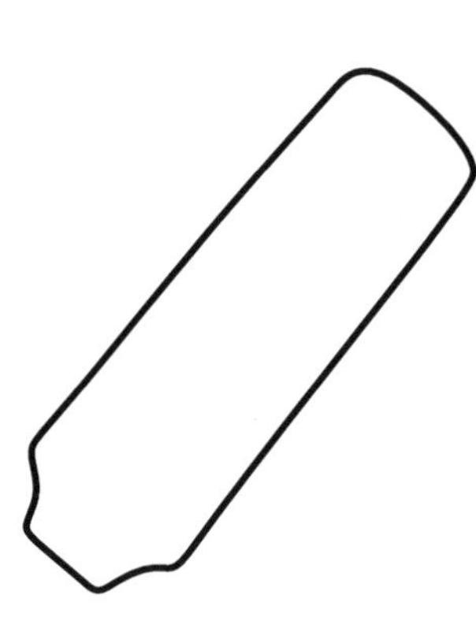

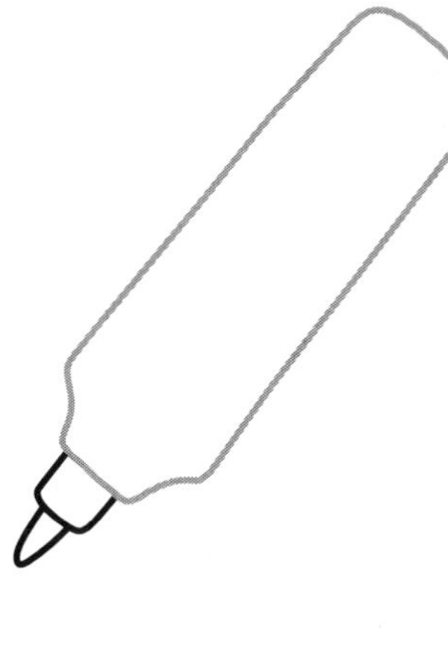

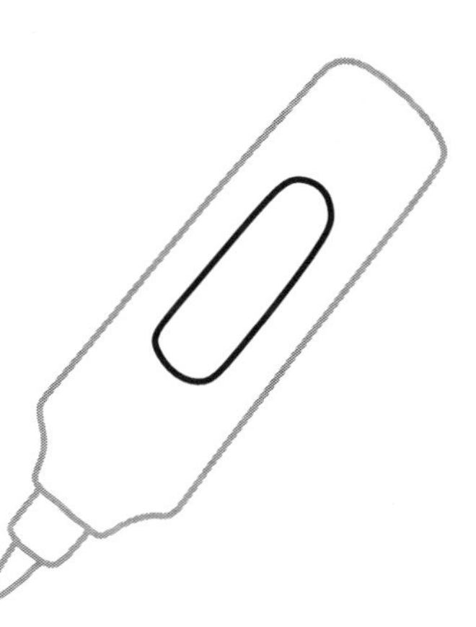

Now, try to draw

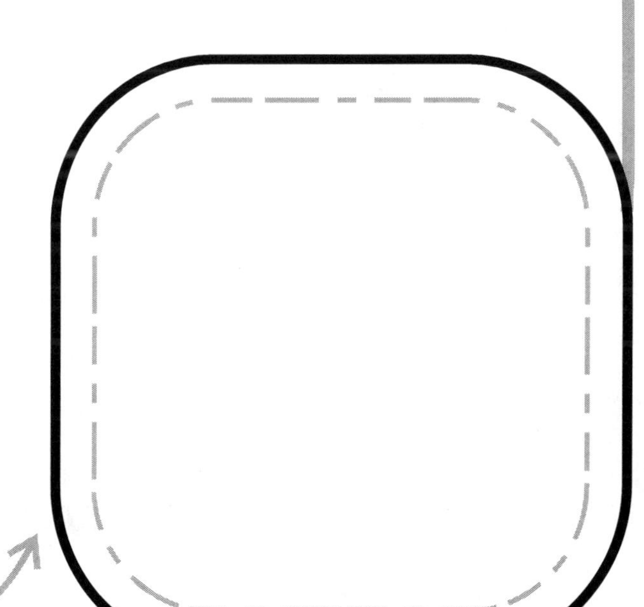

Water gun!

Water guns are fun for outdoor play and soaking friends on a hot day!

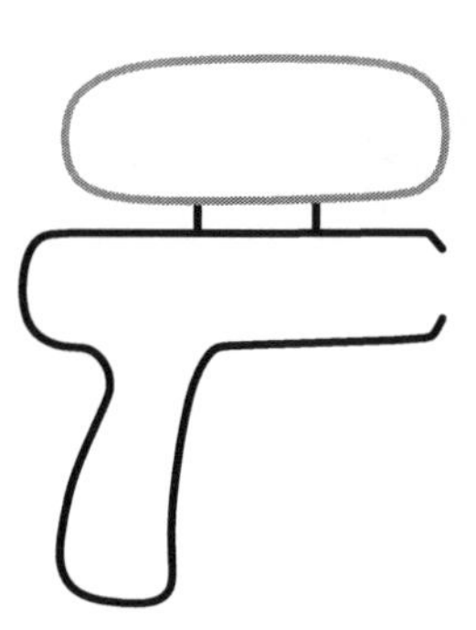

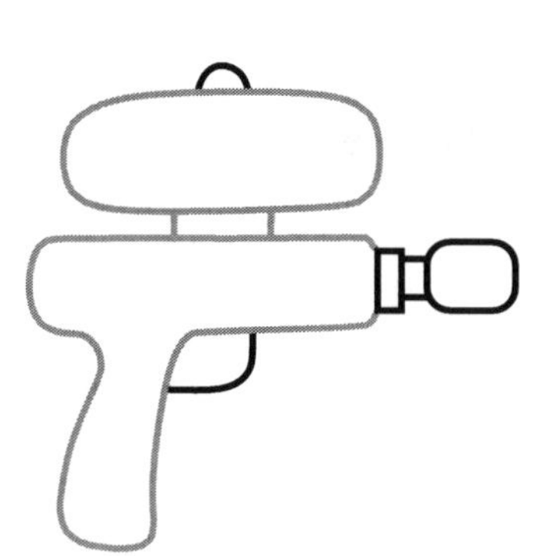

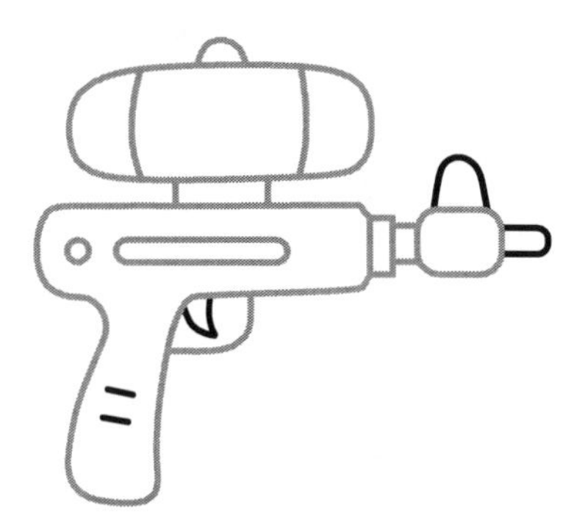

Now, try to draw

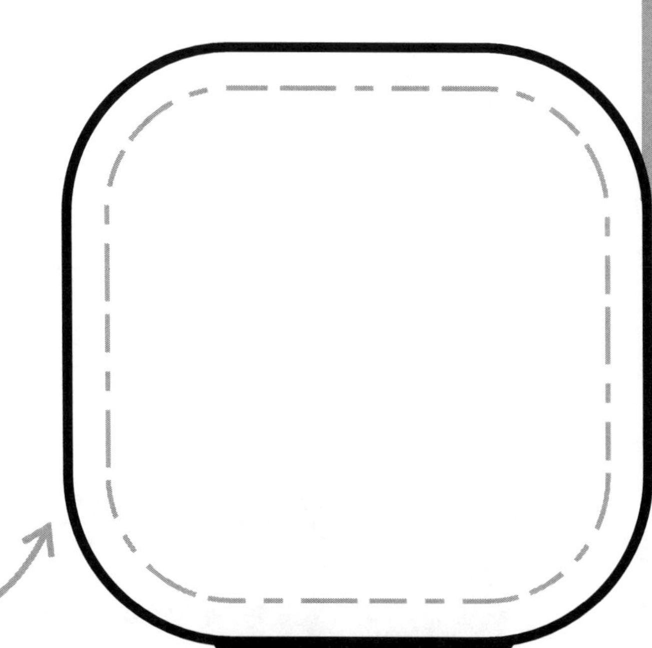

Globe

A globe shows the world, with countries and oceans to explore

1

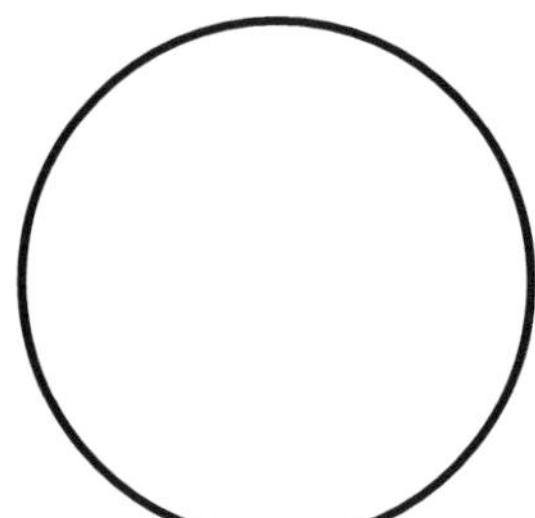

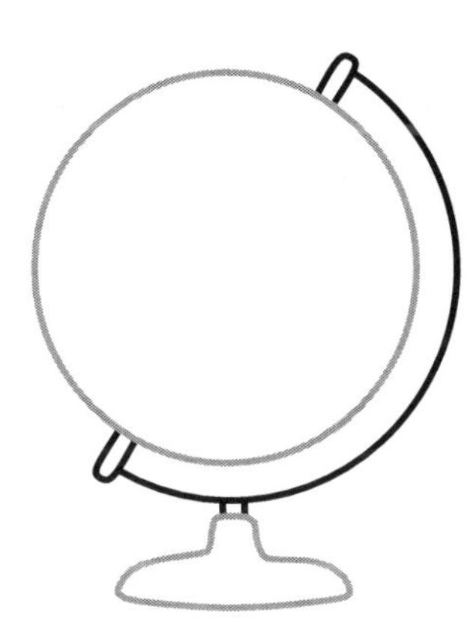

Now, try to draw

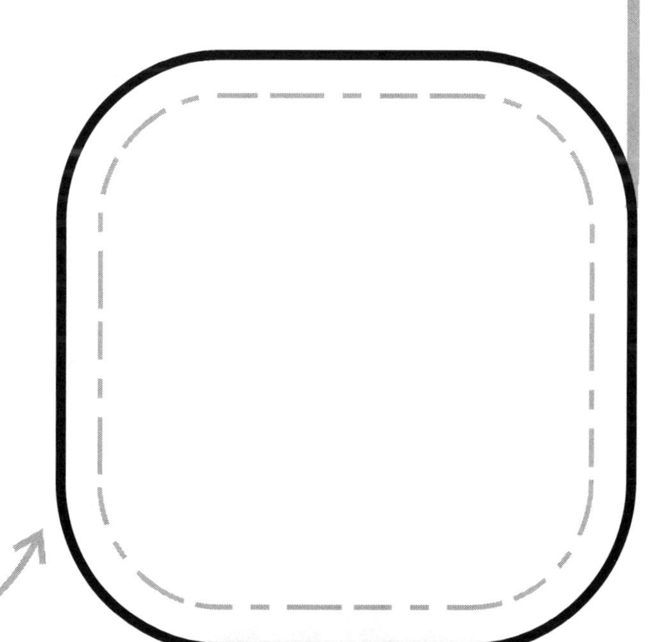

Backpack

Backpacks carry everything you need for school or adventures

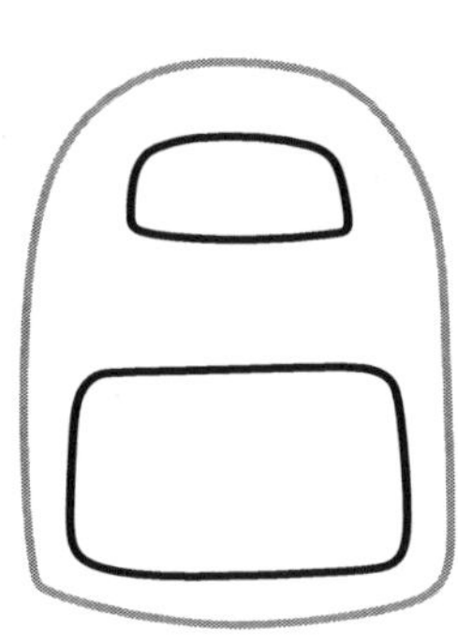

Now, try to draw

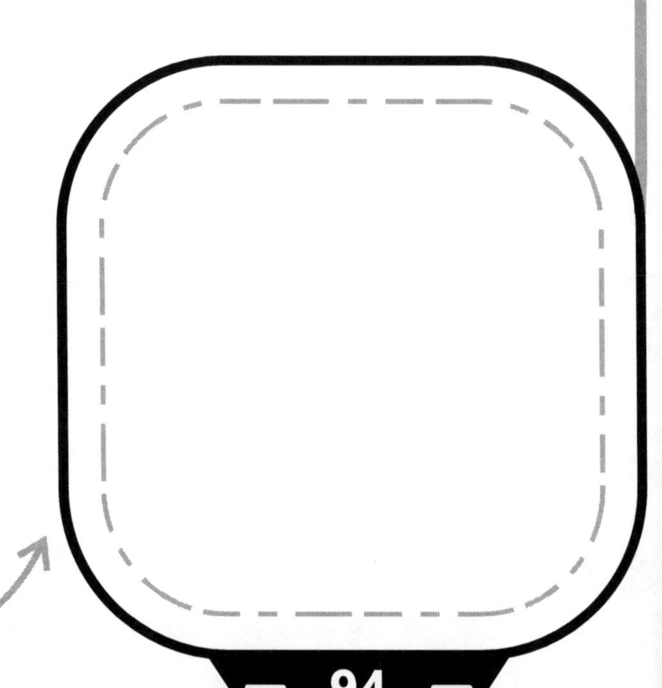

Rocking chairs gently move back and forth, perfect for relaxing

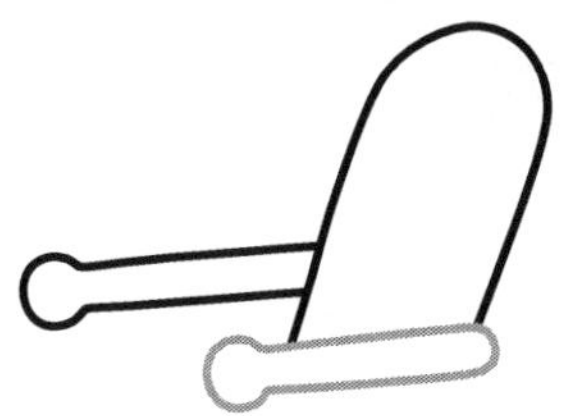

Now, try to draw

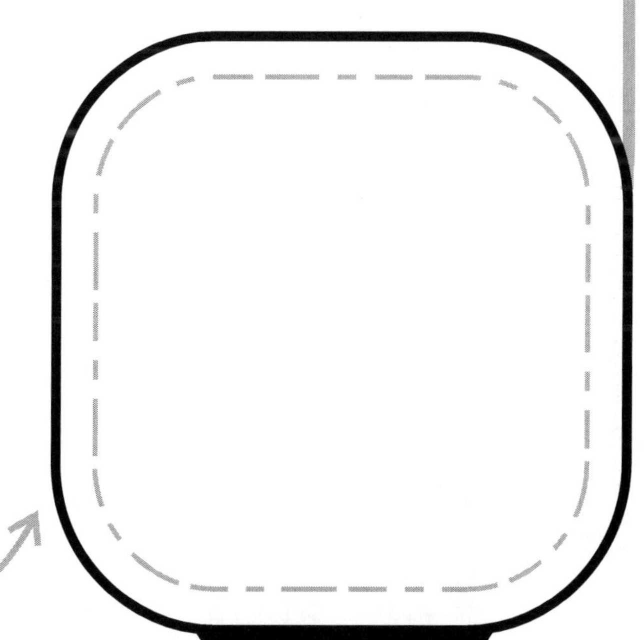

Smartwatch!

Smartwatches can track your steps, show the time, and even receive messages!

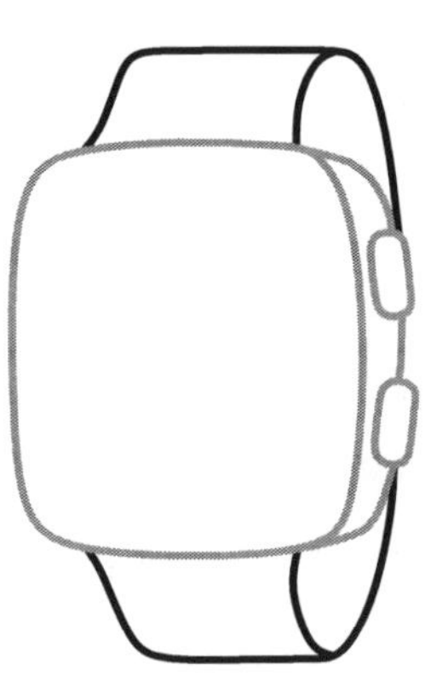

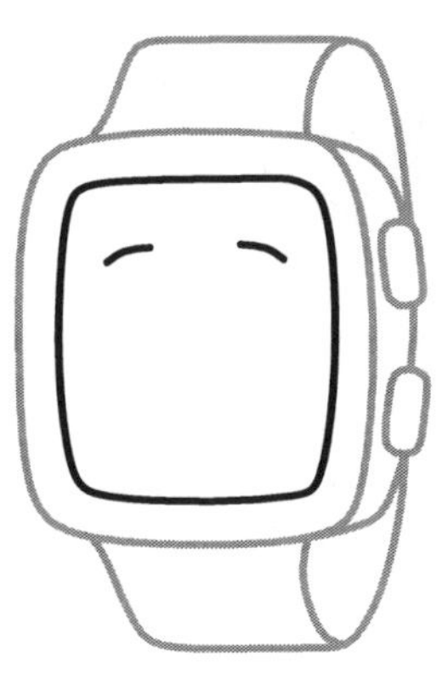

Now, try to draw

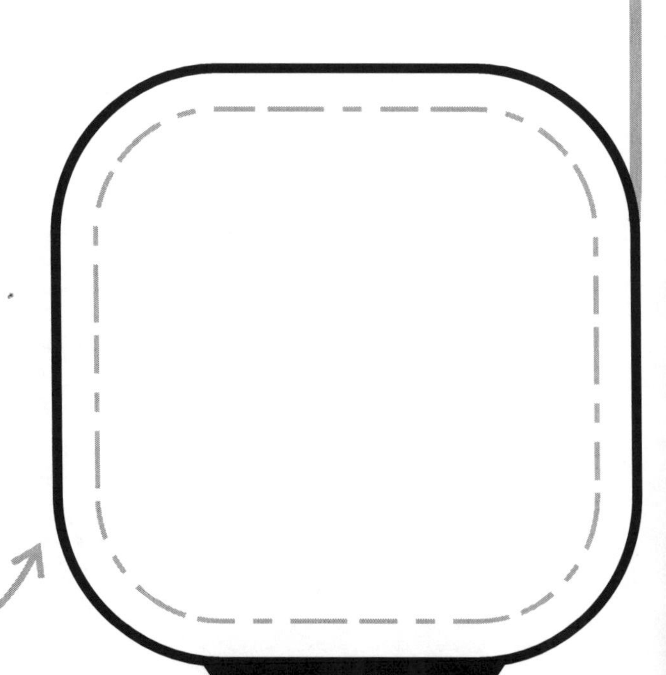

Pizza
Pizza is cheesy and delicious —perfect for sharing with friends!
1
2
3
4
5
Now, try to draw

Burger
Burgers are yummy sandwiches with a juicy meat patty and fun toppings!
1
2
3
4
5
Now, try to draw

Sandwich
Sandwiches are easy to make and can have all your favorite fillings!
1
2
3
4
5
Now, try to draw

Hot dog!

Hot dogs are tasty sausages in a bun—great for picnics!

1

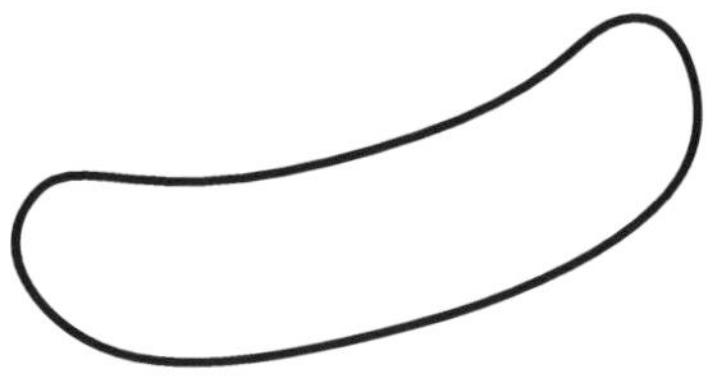

2

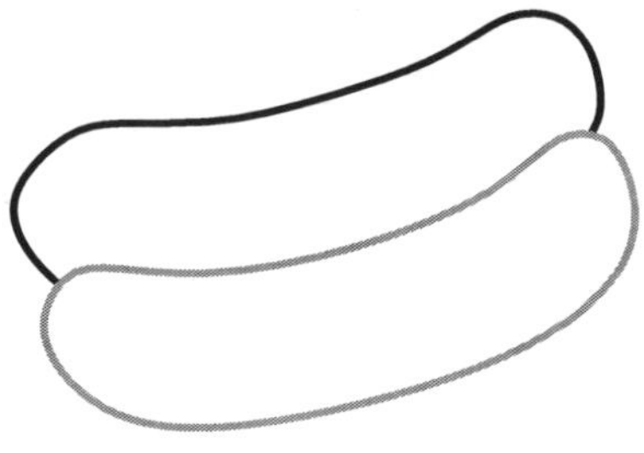

3

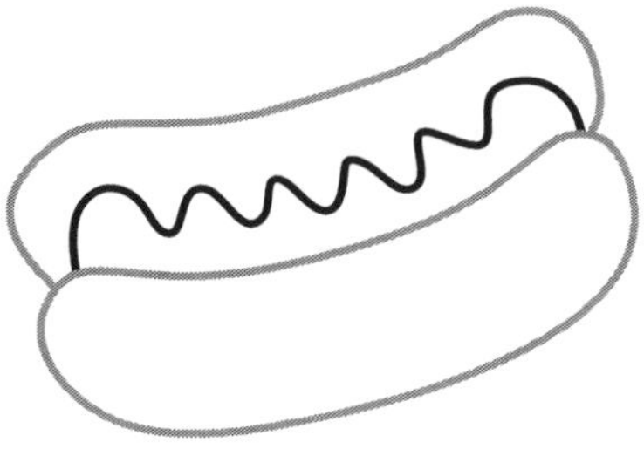

4

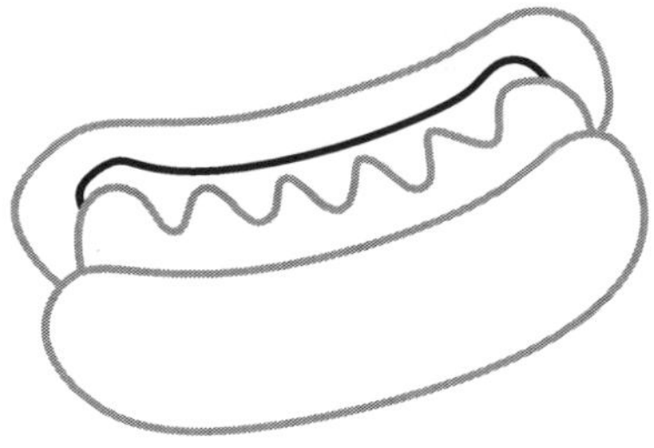

5

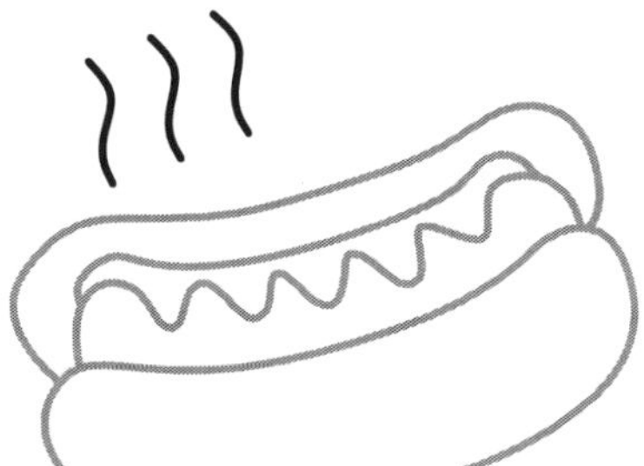

Now, try to draw

French fries

French fries are crispy potato sticks that taste awesome with ketchup!

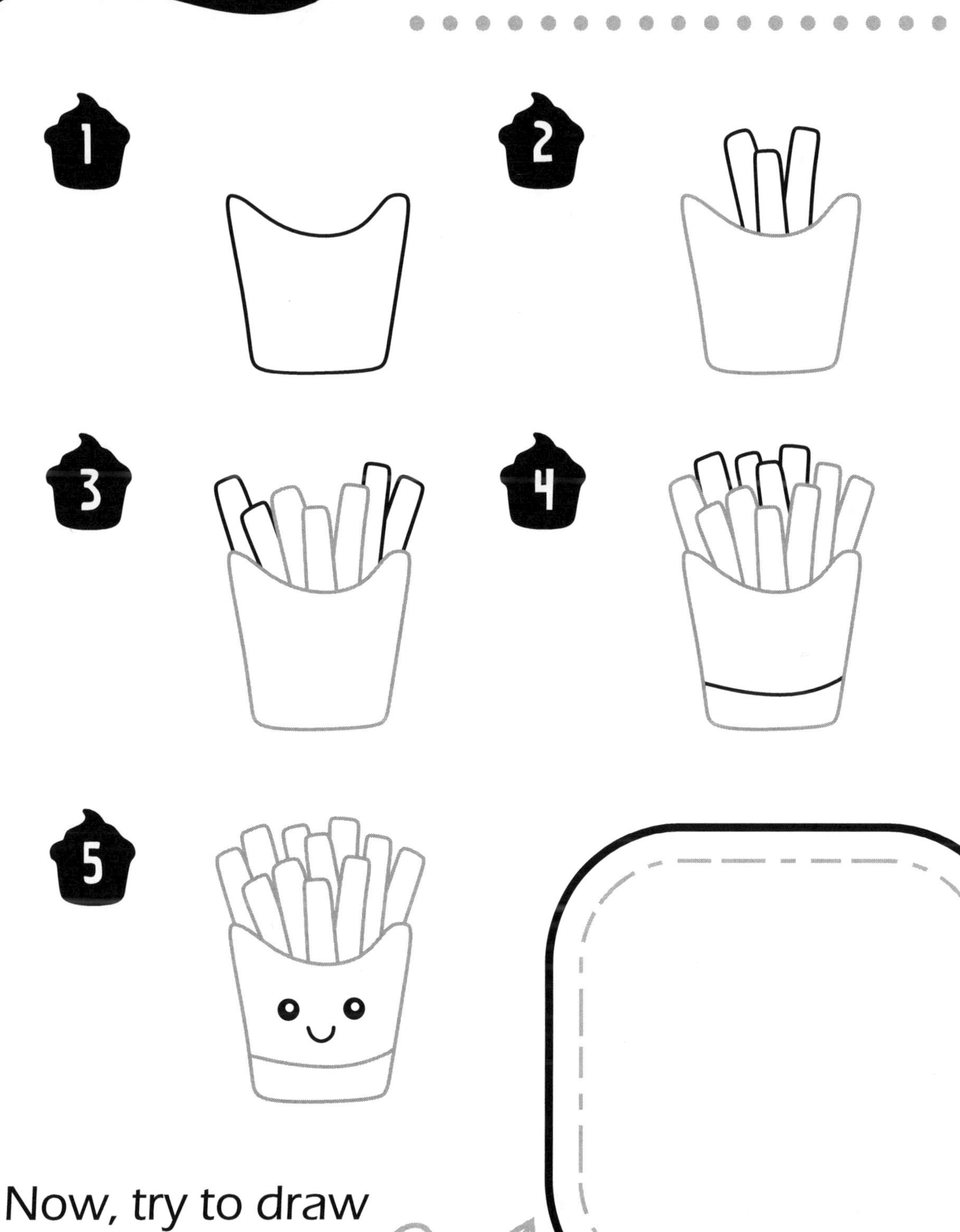

Now, try to draw

Nachos
Nachos are crunchy chips covered in cheese—perfect for snacks!
1
2
3
4
5
Now, try to draw

Lemonade is a sweet drink made from lemons, great for hot days!

Now, try to draw

Apple!

Apples are crunchy fruits that are sweet and good for you!

1

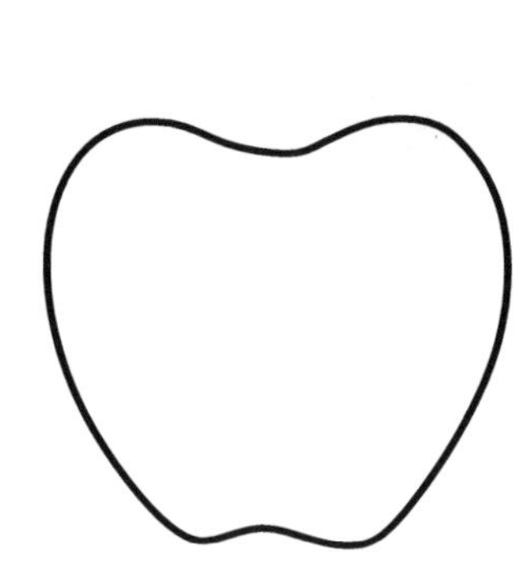

2

3

4

5

Now, try to draw

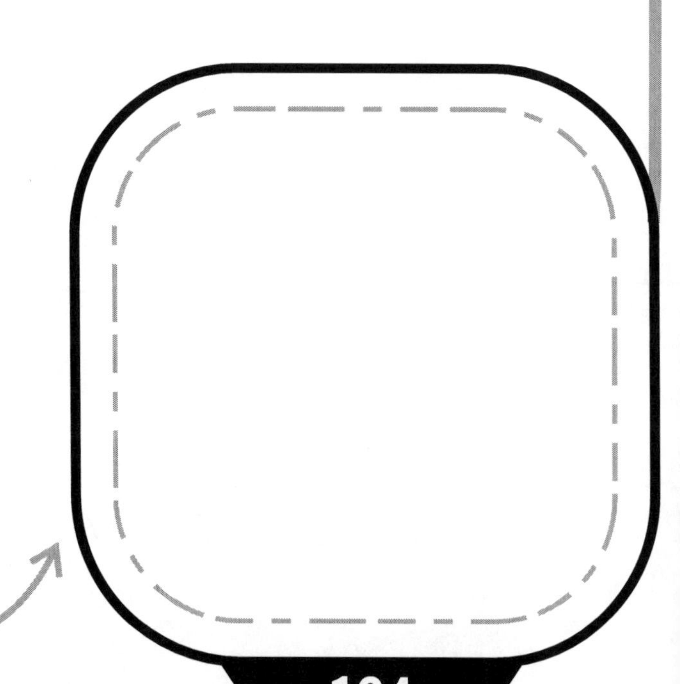

Cake is a yummy dessert often decorated with frosting for celebrations!

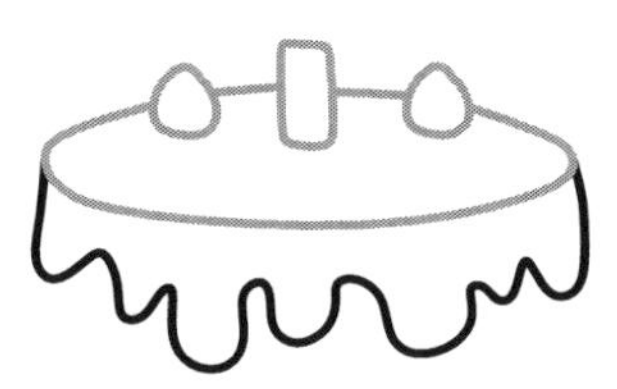

Now, try to draw

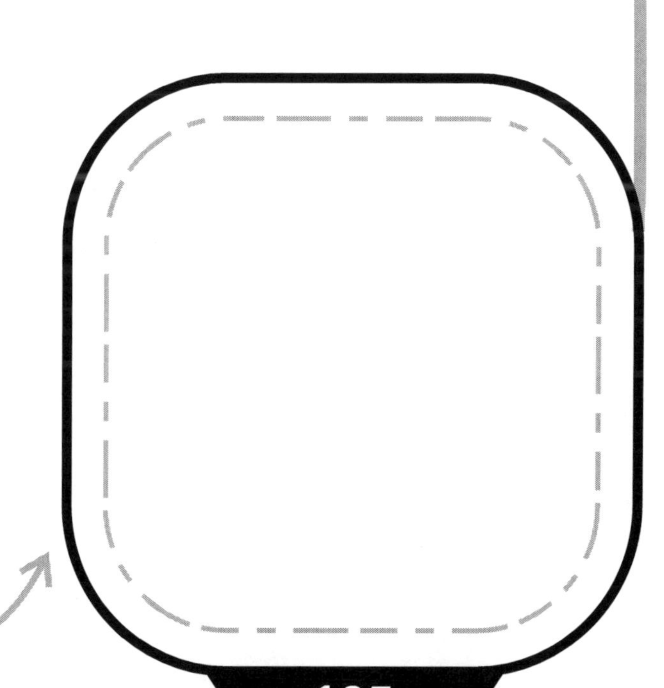

Ice cream
Ice cream is a cold treat that comes in lots of yummy flavors!
1
2
3
4
5
Now, try to draw

Cupcakes are tiny cakes with frosting and sprinkles—so cute and tasty!

2

3

4

5

Now, try to draw

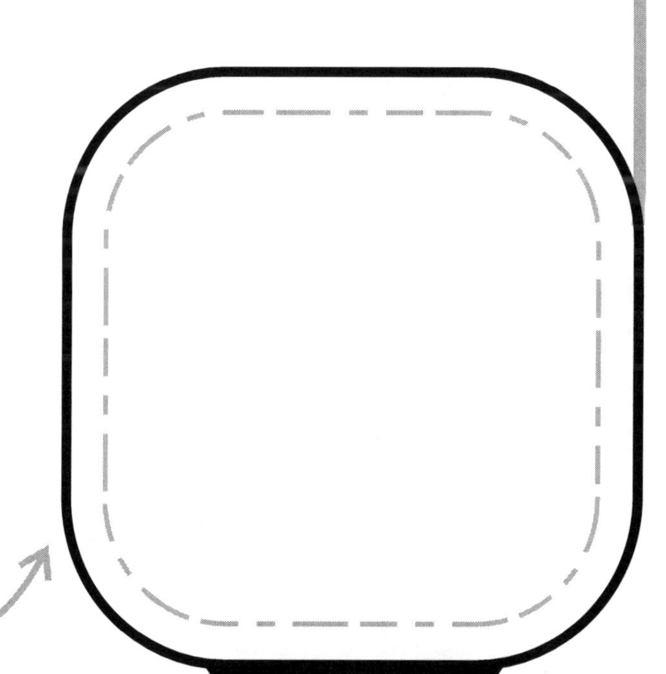

Pancakes are fluffy breakfast treats that you can top with syrup and fruit!

3

5

Now, try to draw

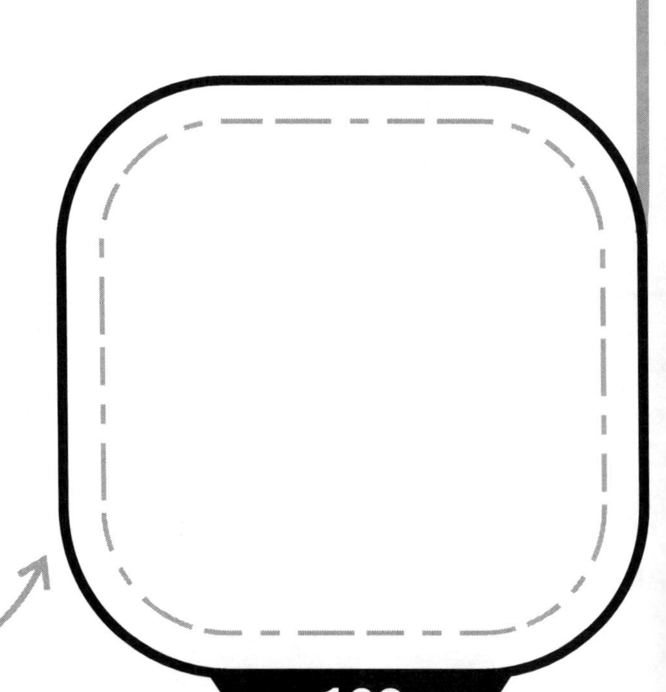

Color Your Own Adventure!
GET CREATIVE AND USE YOUR FAVORITE COLORS
TO BRING THIS THRILLING CAR ADVENTURE TO LIFE!
MAKE IT YOUR OWN!

Made in the USA
Monee, IL
04 December 2024